Also by J D Sleightholme:

Cruising: A Manual for Small Cruiser Sailing
This is Cruising
The Trouble with Cruising
Better Boat Handling
Off Watch with Old Harry

J D Sleightholme
FITTING OUT
PREPARING FOR SEA

ADLARD COLES NAUTICAL
London

Fourth edition 1992
Published by Adlard Coles Nautical
an imprint of A & C Black (Publishers) Ltd
35 Bedford Row, London WC1R 4JH

Copyright © J. D. Sleightholme 1963, 1972, 1977, 1992

First published in Great Britain in 1963
Second edition published by Adlard Coles Ltd 1972
Reprinted 1973
Third edition 1977
Reprinted 1979
Fourth edition published by Adlard Coles Nautical 1992

ISBN 0-7136-3558-4

Apart from any fair dealing for the purposes of research or private study, or criticism or review, as permitted under the Copyright, Designs and Patents Act, 1988, this publication may be reproduced, stored or transmitted, in any form or by any means, only with the prior permission in writing of the publishers, or in the case of reprographic reproduction in accordance with the terms of licences issued by the Copyright Licensing Agency. Inquiries concerning reproduction outside those terms should be sent to the publishers at the address above.

A CIP catalogue record for this book is available from the British Library.

Typeset in Goudy Old Style by ABM Typographics Ltd, Hull.
Printed and bound in Great Britain by Butler & Tanner Ltd, Frome and London.

Contents

	Introduction	vii
1	The work list	1
2	Delivery passage – new and secondhand boats	4
3	An eye on the hull – wooden boats	9
4	Minor repairs to wooden boats – deck coverings	12
5	Glassfibre hulls – repair and maintenance	22
6	Painting wooden craft	38
7	Masts and rigging	52
8	Plumbing, winches, electrics, guardrails	69
9	The auxiliary engine	80
10	Dismantling stubborn fittings	83
11	Bosunry	86
12	Sail repairs	112
13	Safety	121
	Index	129

Introduction

Since a modern yacht is a cocktail of materials – of plastics, synthetics and alloys – there should not really be much that can deteriorate other than through normal wear and tear. Mildew, rot and wood-boring worms certainly cannot touch her. The steady reliability of the diesel engine is taken for granted, and to a large extent so is the electrical equipment, hence a leave-well-alone attitude can arise. The winter lay-up becomes more and more perfunctory and the spring fit-out, which was once a frenzy of scraping, hammering, painting and splicing, is now often a matter of slapping on some antifouling paint, bending on the sails, and stocking up with six-packs. However, wooden boats call for much more attention, and if they do not get it we are soon reminded of the fact by deck leaks followed by decay and an enormous shipwright's bill at the end of the line.

By fitting out your own boat you do a good deal more than save on yard bills. In the course of fitting out you will go over every inch of your boat inside and out, look at every rope, wire and spar, test every working part from winches to seacocks, so that, whether repairs are needed or not, *you* know the condition of everything. On a black night at sea this is either a great source of comfort or a matter for sober reflection and extra care, according to how well you have done your job. Knowing of a potential weakness is half way to avoiding trouble. There is no such thing as a boat being all right so long as we nurse her; she is either fit for sea or she is not.

An extensive fit-out calls for a variety of skills. Very few of us are expert in all of them and because only the very best work is acceptable for seagoing, a wise owner does not hesitate to call in the professional if a job is beyond his abilities. It is the quality of work that matters, not who does it or what it costs. We amateurs cannot always know how to tackle a job; we explore our way into it, whereas the professionals, having coped with similar problems repeatedly, have the experience, the skills and the tools. While it is important to have a go in the interests of learning new skills, the vital thing to keep in sight is the quality of work done. The maxim that 'it will be OK as long as I keep an eye on it' is the rotten rung of the stepladder.

1

The work list

Many people keep a spare page at the back of the log book to enter throughout the sailing season repair jobs that need to be attended to and then cross them out as they are done. The bigger jobs are earmarked for the winter. Thus a jib sheet is renewed, a winch is oiled, but the making and hanging of a new loo door is shelved until the season is over.

The thing to avoid is the hasty lay-up and a closing of the mind to all matters maritime until the following spring, when fitting out is approached with a totally blank mind. It is only when you begin sailing again that you remember the faulty toilet valve, the sticking galley drawer and the shroud plate deck leak. This is the difference between winter work and fitting out. Major jobs are best tackled soon after the boat is laid up and before the weather gets really chilly. Admittedly there may be some calm and sunny days in winter when working on deck is a joy, but down below there will be a deep and grave-like chill and working there will be a misery.

Much depends upon whether a lot needs to be done. An otherwise fit and fully maintained boat is mothballed for the winter and recommissioned in the spring, while others need major winter work and spring refurbishment. If you are likely to need professional help with some aspects, be sure to notify the yard soon after laying up – at any rate do not wait until your wanderlust and the urge to go sailing reasserts itself or your launch date may end up nearer to high summer than springtime.

Organising the work

Most of us are limited to making the best use of what the weekend weather dishes out. In March, April and early May this can mean anything from driving sleet to hot sunshine, hard east winds, to soft drizzle, and if you get your plans wrong you may end up sitting down below unable to get on with work because you have neither the right tools nor the necessary materials for below-deck jobs.

List all the jobs under separate headings: fair weather jobs on deck and foul weather jobs under cover, and then list for each job the tools likely to be needed and the materials required. The best method is to go through each job in your imagination. Take for instance fitting an extra deckhead grab rail in the saloon. The rail will be made at home in advance, so you may need a small plane and chisels for final trimming, drill, bits, countersink, various screwdrivers and so on. Materials will include grit paper, adhesive, screws or bolts, dowel plugs, varnish, brush and white spirit. The rail may be too long so we add saw, G-cramp, bradawl perhaps, and so on. The tools list will be duplicated for other jobs, but we will end up with a bulging tool bag and a mass of vital materials, and nothing we are unlikely to need.

Working parties

Two skilled people working independently can do twice as much work as an unskilled working party which has to be supervised. Unless there is a great deal of unskilled work to be done, such as scrubbing off, rubbing down or slapping on bottom paint, it is usually better to keep your work force small and individual. The problem is that finding enough tools for an unskilled gang to use and then instructing willing workers in their use occupies the valuable time of useful helpers. An unskilled gang can also result in more work. I remember telling a willing novice to give the toerail a light rubbing down and to refrain from using a scraper unless *the old varnish was loose*. He scraped 10 feet of rail to the bone and we had then to scrape 60 feet more of it to match.

If the weather turns foul a large working party will fill the boat down below, leaving you no room to work, while if the weather turns hot its efforts become languid to the point of expiry and of course a long and largely liquid lunch puts paid to any work after 3pm. If a yacht has a crew of regulars either for cruising or racing, I believe it is a good thing to involve them in fitting out. It gives people a stake in the safety of the boat and a chance to put a little of themselves into the general plan. Do be selective though. Find out what skills individuals may have. But at all costs, before you invite a working crew to help, make sure that you have the work, the tools and the materials for them, or at least half the first day will be spent hanging around waiting.

How thorough?

As noted in the introduction, a boat cannot be half fitted out for sea; she must be properly prepared for it or else left in shelter. There is one rider to this, however. Quite often a boat is bought secondhand in one area and delivered around the coast to her new home port. Even when she is a brand new boat she has to be prepared for a sea passage, perhaps out of season, and the tricky question then arises as to how thorough this passage fit-out should be.

The alternative may be to have the boat hauled round by road, but because costs may appear prohibitive to an owner whose cheque book stubs outnumber by far his remaining cheques at that stage of the game, land haulage may be dismissed. What an owner should weigh up is the possible cost of repeated journeys to the boat with overnight or longer accommodation expenses, plus the extra boatyard work and the travel expenses of a crew. Road haulage charges soon begin to look more attractive. More important though is avoiding a passage, maybe in early spring when the weather is uncertain and very cold, with an unknown boat in a guessed-at condition.

It might well be argued that professional yacht delivery skippers are making such passages all the time. However, delivery skippers have a set procedure for checking over a boat, and if they find her lacking they either insist on having the boat put to rights or they may refuse to take her on. It isn't very often that delivery crews deliberately go to sea on a gamble. Bear in mind also that these are very, very experienced seamen, tough and resourceful too. Unless an owner feels he fits this description he would be wise to think twice about a winter or early spring sea passage, which can be utterly taxing both to boat and human stamina.

Assuming though that a delivery passage is considered to be a sound option, it is less a question of the amount of work the boat needs, but more how little we can safely get away with. The varnish work may be badly neglected after a couple of years of standing out in the elements, but this is irrelevant in terms of safety, whereas a fuel tank containing stale diesel, condensation and sludge most certainly is relevant. Chapter 2 should be read in the full understanding that it is not a regular fitting-out programme to be followed annually; its aim is to enable a boat to make a single sea passage of short or moderate length. It will not guarantee that nothing will go wrong. When a boat is new to us there may be many little secrets that are only discovered by sailing her. On arrival at her new home port the full fitting-out routine must be put in hand.

2
Delivery passage – new and secondhand boats

The new boat

In the case of a brand new boat there should, in theory, be nothing to go wrong. Everything will be in pristine condition, but by the same token nothing has been tested either. If a boat is the first of a new class there may also be design faults that come to light and which will be modified in later marks of the new class. It is by no means unusual for a brand new boat to reveal problems on her first sail, which of course is why big ships always have acceptance trials.

By way of example, one new boat I owned had the exhaust pipe fitted contrary to the engine manufacturer's recommendations and I finished up with an engine full of seawater on the first run. Another boat shed her rudder after her first six months of life afloat due to a wrong mix of metals and the resultant electrolysis. In yet a third case of a new boat on her first passage – a vessel that had electric fresh water pumps everywhere – all pumps failed at once and we finished the passage on canned ale and with stubbly chins.

The moral when planning a passage in a new boat is to take nothing for granted, carry a big bag of tools, and make every effort to have a short day's sail in her before taking off into the blue. I personally cannot ever remember doing this – I was always in a hurry to make the best use of fine weather or impatient to be off, but it is sound advice just the same.

Checklist for a new boat

Even when a boat is brand new she must be thoroughly checked out in every working aspect. The engine should be given a run under full load both ahead and astern and the batteries should be verified as receiving a charge.

- Check seacocks
- Test all lights
- Check rigging and mast tune
- Check length of anchor cable and secure its end below
- Test bilge pump
- Check gas installations for leakage (by means of liquid soap on joints and unions)
- Fill water tanks and taste, check pumps
- Hoist sails and check sheet leads, etc

Early in the passage check the log, echo sounder and compass. Use buoy-to-buoy runs to check the distance run by log – this is a rough test only and proper log calibration must come later. Compare echo sounder readings with chart notation adjusted for height of tide, and compare compass headings with bearings taken by hand compass on the yacht's fore-and-aft line. These will be rough headings and the compass will have to be adjusted if necessary later on. The purpose of these checks is not so much to ascertain the accuracy of the instruments as to discover any gross errors that may be present, to identify inaccuracies and, in the case of the log, to establish any errors in terms of percentages (eg under-reading by 20 per cent or whatever). It is by no means unusual to find a new boat fitted to the owner's instructions with electronics of various sorts which have stacked up massive compass deviations.

The secondhand boat

The delivery fit-out of a secondhand boat will be closely linked to her age, condition and general history of maintenance, plus the time she has been laid up awaiting a buyer and whether she has been stored under cover. Thus a boat might be taken over by a new owner from a careful previous owner in full seagoing condition and in midseason. Conversely, she may have been dumped on the market, perhaps following the death of her owner, and left standing neglected and unprotected in some boatyard for several years.

This sort of knowledge of the facts is vital. Years of skimped maintenance may have brought a boat to the point where a dozen and one things are about to go wrong. The old no-maintenance myth dies hard, and many owners think that laying a boat up for the winter means no more than locking the hatches and walking away. Thus a new owner inherits a boat with flat batteries, corroded electrics, seized seacocks, dirty fuel and engine filters, clogged-up bilges and a mast that has been standing uninspected for years.

If a boat has been professionally surveyed as part of the process of buying her – as indeed she should have been – there will be a list of observations and recommendations to guide her new owner; it should always be remembered, though, that a survey of a boat ashore and out of commission can only be an assessment of her general state of health, not her working fitness. Her engine needs a separate survey running under load, and the condition of her rig and sails in a hard breeze remains to be discovered.

Checklist for a secondhand boat

If the boat has been laid up in the yard for some years, ask the yard foreman what work has been carried out on her and whether she has needed any drastic repairs. Tact may be needed though, because the previous owner may still be a valued client who is buying another boat which is to be kept in the vicinity. However, it may be possible to find out if the boat has a history of mechanical ailments or electrical problems, whether rigging has been renewed, and so on. This sort of friendly probing into the background of a boat you have just bought is reasonable enough in any case.

Chatting to yard engineers, shipwrights and riggers in a casual fashion can sometimes uncover valuable information, and if a yardhand rolls his eye and grins at the mention of the late owner's name you can make a mental note that nothing must be left to chance – and buy the man a drink!

Engine   Was it properly laid up, winterised, oil and filter changed?
Seacocks   Check that all are working freely.
Lights   Check that all are working. If masthead lights are not visible in daylight, either go aloft or check after dark – this is vital.
Battery   Have a test carried out to determine battery health, especially if they have been left in the boat and lying idle for a long spell.
Rig   Examine every wire, tang, shackle and rigging screw closely and check that rigging screws are seized.
Guardrail wires and *stanchion feet*   These need to be tested.
Gas   Check the installation.
Check other details as for a new boat, but pay particular attention to the power sources, ie *sails* and *engine*.

If the mast is standing, either go aloft or pay a rigger to carry out a careful check. Masthead sheaves may be jamming, there may be corrosion caused by the unwise mixing of metals (self-tapping screws of the wrong stainless alloy for instance), distorted shackles, or bent pins.

If the standing rigging is of stainless steel wire (as is usually the case) and if the rig is over twelve years old, it is already a bit suspect – especially if the boat has been raced hard and the wire is of minimum acceptable size in the interest of weight-saving. Look at the crosstrees carefully for signs of kinking or bending. An earlier brush with another yacht can do damage unnoticed at the time, but which becomes a potential hazard later leading to the loss of the mast.

Lubricate the mast track or luff groove, and check the roller headsail (if fitted) for free rotation with no grating or 'lumpy' patches. Once the main shrouds and stays are set up (see references to this later), seize rigging screws if not already done. Even a seizing of thick sail twine will serve for the duration of the passage – don't rely on the locknuts. Lee wires hang slack when sailing hard and they are subject to movement which can cause screws to walk back.

Be suspicious of guardrail lashings. Synthetic fibres that are not UV proofed can become 'sun-rotted' to the point where they may snap like kitchen string. The previous owner may have yielded to temptation and bought some 'bargain offer' lashing stuff.

Don't take the bilge pump for granted; haul up the suction pipe and make sure that it is clear and that it has a filter of some sort on it. Throw a few buckets of water into the bilge and test the pump – for all you know, the sternshaft gland may be leaky. Check also all freshwater tanks, pipes, pumps and connections; taste the water and treat (see later comments) if necessary.

Spend time trying out the options for sail reduction: mainsail reefing; means of reducing the fore triangle; and if a storm jib is part of the inventory, how this is rigged. All too often we find yachts that carry a storm jib that has never been used and which, in any case, may prove impossible to set and sheet home. If the hard weather option is supposed to include a deeply rolled headsail, be suspicious of its value in a real blow and take a close look at the roller line which will come under heavy strain when the jib is half rolled in heavy weather.

With a delivery fit-out we are only concerned with getting the boat home without anything going seriously amiss, but engine reliability is essential. If it has been lying idle without having been properly laid up, it would be wise to get a marine engineer to check it over. An engineer has a trained ear for trouble. If, on the other hand, the engine was correctly laid up, all filters have been renewed, the oil changed, the pump impeller checked and/or changed, and if the engine was running smoothly prior to lay-up, there shouldn't be much to worry about. If the boat is out of water and a hosepipe or water container and funnel can be rigged, the engine can at least be started up; but it should be understood that not until the engine can be run under full load can we be sure that it can be relied on. Starting up while ashore will determine whether the engine is firing smoothly and whether the alternator is charging the battery.

If a boat has been laid up for several years with fuel in her tank it is best to pump or siphon out every last drop and replace it with fresh fuel. You may well be horrified to see the state of the final dribble of old fuel – an emulsified sludge that would otherwise find its way into the system as soon as the yacht hit rough water.

If the engine is satisfactory, the rig is in sound condition, lights work and you know where to find the fuse-box, safety equipment is serviceable and navigation instruments are basically functional, you have only to worry about hull integrity – hence the need to ensure that all seacocks can be turned off if necessary. If any do not work, don't force them. If the application of penetrating oil and reasonable pressure fail to shift them, either get them fixed or replaced before leaving or plug them with soft wood pegs from outside. Wheel valves snap at the stem very easily if forced.

3

An eye on the hull – wooden boats

Fitting out concerns a boat's seaworthiness; hull condition is more important than mere appearance. In a wooden hull any weakness will probably be the result of wood rot, unreliable fastenings will possibly be due to old age (and consequently movement of the hull planking with attendant leakage from loosened caulking); marine worm borers in certain cases; cracked frames, split planking and plain neglect.

Where wood rot is the source of trouble it is most likely to be a result of leaking decks initially, and closed, unventilated spaces secondarily. Moisture penetrating a faulty deck seam or joint around the coachroof, coaming, hatch, etc can travel along the space between two members, such as a deck beam and the deck, for considerable distances before the leak is detected below. In its wake it leaves saturated timber in airless spaces; this is the perfect breeding ground for fungal spores to colonise.

Planked decks that have become nail-sick, and therefore slack, will not hold the caulking in the seams; a common remedy, rather than costly refastening, is to cover them with canvas or plastic sheet, or alternatively to coat them in heavy deck paint. There is no real cure for a nail-sick deck. When an ageing wooden boat comes on the market and her decks show signs of repeated painting, temper enthusiasm with caution.

In the case of a planked and unpainted deck, look at it carefully as it is drying off after a shower of rain or after washing down – quite often the last seams to dry will be those that are faulty and retaining moisture. Take a sharp pricker and attempt to lift the seam filler and caulking; if it lifts easily the whole seam may have to be raked out, allowed to dry thoroughly, then recaulked and paid with seam filler. Use either the synthetic type from a tube or the old-fashioned 'glue'; the latter is a kind of pitch and is poured in a molten state, the excess being scraped off later when cold.

Check areas around fittings, pads, beadings and so on by using the pricker; soft wood may merely be wet wood in need of attention but thus far unharmed by rot. Force the point in and withdraw it; if the wood is rotten there will be no resistance to withdrawing. Lift a splinter of wood with the tip of the pricker – good wood will be fibrous, affected wood will snap short.

If a leak is found on deck, trace it below; conversely, a leak below must be traced back to its origin on deck. Sometimes a harmless dye can be used, such as cochineal or permangamate. The search may mean having to remove deckhead liners or side panelling, both of which are bad news in wooden boats because of the dead air spaces they provide; so now may be the time for a total rip-out, replacing panels with plain battens or exposing the bare beams and timbers completely. After all, there is no hardship in having to gaze upon a shipwright's handiwork. Below decks there should be no dry joints to attract and harbour moisture; get rid of linoleum – this is a notorious cause of rot.

Wood rot can either be wet rot, as described, or dry rot. Dry rot is less common but is a more virulent form that spreads rapidly – it is so contagious that the spores can be carried by hand. In both cases, rotted wood must be cut right back and well into sound timber and the adjoining timber; replacement timber must be deeply impregnated with a chemical fungicide. No repair is effective unless the cause is dealt with and the whole area is opened up to ventilation; rot spores are like a deadly cancer and the smallest remaining affected area will spread deep into the grain of sound timber if it is not eradicated.

Hull planking may appear sound yet the frame behind be infected; in this case, the fastenings will soon loosen and the boat will become leaky. A surveyor armed with an ebonite mallet can detect hidden rot by tapping along the planking where it is fastened to frames; a dull note or a firm ringing tone reveal the condition.

Usually deck and other leaks are betrayed by the discoloured dribbles and streaks visible below and rot can be detected by discolouration of timber or the buckling and waviness of painted surfaces. When rust streaks are seen, however, the problem may be a sign of bolts, spikes or such fastenings corroding. The problem then arises as to whether to replace these – often a highly difficult and costly operation. Although yachts were seldom built using iron or steel bolts and fastenings, ex-fishing boats certainly were. The extra problem of electrolysis caused by mixed metals in a salt water environment threatens all fastenings including bronze, stainless steel and the rest. Keel bolts are a classic source of worry in an ageing wooden hull, and every owner should know the general condition of such vital fastenings either by having a bolt withdrawn for inspection or by calling in a mobile X-ray operator.

Plywood-built boats

Although plywood has fewer seams and joints when used for hull construction and marine grades of ply are virtually indestructible by moisture, ply is an almost ideal territory for the breeding of fungal decay. Once rot is allowed to attack the open end grain of a plywood sheet it can run like wildfire beneath the surfaces and can go undetected until the tell-tale waviness of the surface begins to show. I once saw a sailing dinghy that was out of control hit a large plywood hull amidships and go right through her topsides for nearly 2 feet of her bows.

Thus with a plywood hull it is the edges of the sheets that constitute the risk; look for signs that paint is not adhering properly, or soggy seams, buckling edges and, as in the case of deck seams, look for areas that don't dry off as quickly as the rest after being wetted. A plywood hull built of marine grade timber and scrupulously maintained can literally last a lifetime – or conversely, it can be a write-off in a few years. There is no treatment for decaying plywood other than wholesale replacement. Dabbing on fungicide is about as effective as treating a broken leg with iodine.

4

Minor repairs to wooden boats – deck coverings

In the context of fitting out, repairs are strictly cosmetic; any major shipwrighting work should be put in hand during the winter months. The canvasing of decks, however, being a fine-weather job, may well have to be left until early summer if the boat is out of doors, thus forcing a delayed fit-out. Without deck integrity, of course, a wooden boat has embarked upon a steady period of ruin.

Graving pieces and fillets

The odd bangs and bashes of the summer sailing season usually cause very minor damage – deep scores or damaged plank edges, or possibly a deeply dented stem if it has been a head-on confrontation. None of these wounds are serious, but they are too deep or too awkward to be disguised with filler or paint alone. The proper answer is the carefully fitted graving piece or fillet.

First remove all paint and varnish to bare wood so that the grain can be studied carefully. A graving piece will consist of a shallow but precisely cut slot in the surface of the damaged plank or spar, deep enough to provide a secure bed for the graving piece, which will be cut exactly to fit and then glued in place; it must not be nailed or screwed as it is made slightly oversize so that the whole job can be planed flush later. Cut the slot with its ends and sides exactly square and clean. If possible, let the width of a suitable chisel dictate the width of the slot. In some cases it is better to cut the slot in the form of a diamond as this minimises the amount of end grain glueing to be done (Fig 1). Either way, the important thing is to make a good fit which will not allow moisture to seep underneath.

Fillets may take the form of a new edge to a plank which has lost its caulking V or it may be a means of repairing a longitudinal crack. Again, in the case of a seam, a complete 'spline' may be fitted between the two plank edges in lieu of normal caulking. It will, of

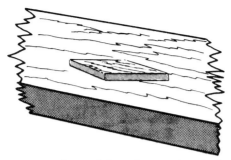

Fig 1. Graving pieces: a diamond slot reduces end grain glueing.

course, be glued in and planed flush afterwards.

Where a crack in a plank is to be tackled we have to assume that the crack runs clean and straight along the grain. If it does not and if it consists of a number of linked cracks, perhaps following a diagonal line of grain, it would be better to renew a section of plank. With a simple straight crack, the first thing to do is mark out a slot perhaps ½ inch wide and running a little beyond each end of the crack (Fig 2). This is then sawn out carefully. The sides of the slot are now

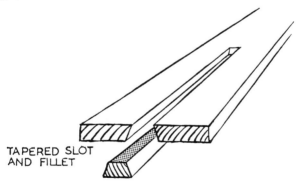

Fig 2. Graving fillet.

given a slight taper and the ends similarly treated. A fillet is cut to fit, tapers and all, and this is glued in place. Once more it should be slightly thicker than the plank to allow for final planing flush.

In the case of a damaged stem there must be a careful inspection within the hull to ensure that the bash has not jarred the stem away from its hook scarf joint to the keel and other components. There will be a disturbed paint line perhaps, or evidence of slight movement elsewhere. In such cases it is as well to take professional advice as the damage may be more serious than supposed.

With a hinged centreplate to chip and paint, the yacht must either be held suspended on a crane or a hole must be dug below the keel into which the centreplate can be lowered.

Assuming that the bash has only dented the sharp edge of the stem and that the damaged wood can be cut away without weakening it, saw out a slot which is long enough to give the new fillet sufficient glueing surface along its length and cut the ends in a slow taper. The new piece, probably of oak, can be left well oversize to allow scope for shaping up afterwards. Screws may be used but care should be taken to choose a material similar to the rest of the fastenings.

Choice of fastenings

As just mentioned, the material used for fastenings must be compatible with the rest of the metal in the hull. Brass screws have only a

Hull has been burned off, sanded flat, and all hollows, scratches and indentations have been filled with trowel cement.

limited life in any case because the alloy wars within itself when subjected to salt water and the metal turns reddish in colour and becomes brittle.

Galvanised wire nails are suitable for use on an iron fastened hull – usually an ex-fishing craft, although hammering destroys the galvanising on the head and rust weeps will result. Various galvanising paints are available that can be daubed on, but this only provides a limited cure.

Gripfast nails of monel metal are expensive, but worth using for limited repairs; once driven though there isn't much hope of being able to draw them out again because of the saw-tooth rings which give them their one-way grip. Copper boat nails, square in section, must always be drilled for as they are too soft to be driven and they should never be used in company with iron or steel fastenings.

The glue can be a recommended gap-filling type such as the urea-formaldehyde and epoxy ranges. There are a number of such adhesives, but some require a controlled temperature and are unsuitable for open air work, especially in cold weather. Even the best of glues will prove disappointing if wrongly used. Every glue has its setting time and 'shuffling' time during which the work can be moved (as when tightening screws etc). Failure to keep within the prescribed shuffling time will destroy the bond completely.

Decks

As noted earlier, a leaky deck is a sure way to allow rot to gain a hold below decks in a wooden hull. A planked and caulked deck will, in time, become leaky as a result of the age of the vessel and the tendency for fastenings to slacken and thus to allow movement to take place.

A tight deck depends upon the planks being firmly fastened and with seams of regular width which retain that essential V. Even a deck in good order will develop leaks though, and it is advisable to go very carefully indeed when attempting to caulk an otherwise sound deck. If the caulking cotton is hardened in too much in one spot the whole tension of the seam will be upset, just as in driving in a row of wedges – if one is driven harder than the rest it causes them to become loose.

An obvious seam fault should be examined. The stopping compound may be faulty and by removing it, allowing the seam to dry and then re-stopping it, the fault may be cured. Likewise, as mentioned earlier, a leak coming from below a deck fitting may be easy to trace and cure in the same way.

Old decks with wide seams that won't hold their caulking must be renailed where necessary and then covered either by applying one of the many special mastic compounds or by canvasing or glassfibre sheathing. First, though, the deck must be in good repair and free from soft patches, it must be planed level and scraped bare of varnish, etc. Oil patches must be treated with a degreaser. All fittings, beadings and hatch coamings must first be removed, of course, and the rubbing strake removed from the deck edge, also the toerail or bulwarks (Fig 3).

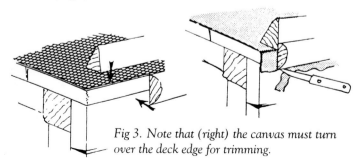

Fig 3. Note that (right) the canvas must turn over the deck edge for trimming.

After allowing the decks to dry thoroughly (and it may be very damp under the toerail and other long hidden places) the seams must be carefully stopped and levelled. If a mastic or liquid plastic is used it may be safer to use a white lead stopper as, although this will harden, it is less likely to set up conflict with the mastic than some of the synthetic stoppings. At any rate, read the mastic instructions carefully and if a special stopping and undercoat are advised be sure to use them. The mastic may be trowel applied and it is important to build up a thick even coating over the whole deck.

There are two schools of thought on laying deck canvas. It can either be laid on a thick coating of paint or glue or soaked with salt water and stretched over an already painted deck and wetted again before the canvas is painted. The first plan, in theory, sticks the canvas to the deck, thus prohibiting any movement. If there is movement though, it is likely to cause the canvas to stand proud in ridges. The other plan makes it possible for the deck to move if it wants to. Of the two I prefer the wet paint or glue method as this prevents any chance of moisture lying between canvas or deck and the deck should not move that much. A damp-trap is lethal to a wooden deck.

What is most important is that the canvas be stretched evenly while being laid. I think it is uneven stretching that later causes the worst wrinkles. First, though, the canvas must be planned, cut and sewn together. Assuming it to be 36 inches wide (it may be narrower), an 8 foot beam will call for two widths plus two narrow strips in the region of the sidedecks – and incidentally the greatest beam. A good plan is to make a full-scale pattern using heavyweight wallpaper, cutting it to fit exactly round hatch openings and the coach roof coamings – there will have to be an allowance made for turnovers, but this will be done when the actual canvas is being marked out.

The difficulty is that there will be some stretch and if the canvas is sewn up and cut roughly to shape around the edges, allowing for the deck edge turnover, it can then be pulled out evenly over the paper template to see just how much stretch will occur during laying. Remember though that once the deck openings are cut, the tension will be altered and the narrower parts will tend to stretch more than the full widths. The coachroof and sidedecks may be best tackled separately and such openings as the forehatch (coamings removed) may be covered over and cut out later.

Some people prefer to tack the canvas down along the centreline and stretch it out towards the deck edges; others favour securing it along, say, the forward edge of the coachroof and working it forward. The coachroof itself may already be canvased, otherwise the method is the same as for the deck.

Be sure that the warp of the canvas runs fore and aft and before taking the cut strips to the sailmaker for machining make sure that there is enough allowance for the seam. The edge of the canvas will have a selvedge and will not require a turned over seam; this is important as the seam must be as flat as possible. The weight of canvas used is less important than one might suppose and an 8 ounce duck is suitable. See that all corners are neatly mitred, and when laying the canvas on wet paint draw it out and smooth it down to expel any trapped air. Use copper tacks around hatch openings, spaced at about 1 inch and around the deck edge, where excess canvas may form bulky undulations when it is turned down to cover the deck edge, cut a series of V nicks. These edges can be laid over a thick luting of paint and putty mixture and tacked down neatly. The rubbing strake will cover them later. Hatch coamings and beadings will also need a thick luting upon which they can be screwed down. Give the canvas a coat of paint before doing this though and finish with a further four or five coats, later adding a last coat of non-slip deck paint.

Laying plastic deck coverings

The heavy plastic embossed-pattern coverings have largely replaced canvas and, while the preparation of decks, removal of fittings and structures and so on is the same as for canvasing, the rest of the operation differs considerably.

These coverings must not be stretched when laying, therefore each section will have to be measured and cut with great precision and thick paper or card templates are essential. Remember too when planning the various sections that the material has only one 'right side up', therefore the cutting out of pairs of left and right sections must be planned carefully. See also that the pattern of dots embossed on the surface runs consistently in each piece; one small section of decks having, say, diagonal lines of dots will stick out like a sore thumb if the other dots run square fore and aft. The accompanying sketch (Fig 4) shows how the sections of deck coverings should be planned.

The midship line of deck should be pencilled in and the decks must be absolutely free from grease, tar, etc. Bare wood is preferable. When all the sections are cut and have been laid down in a dummy

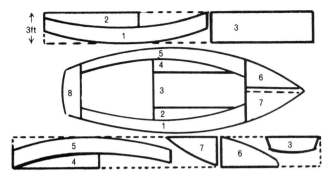

Fig 4. Templates for plastic deck coverings.

run to ensure that they will fit, remove them and plan out how each piece is to be handled when sticking down. The adhesive provided makes instant bond and once a piece is down it cannot be taken up and relaid. The aim is to start lowering each piece from a predetermined position so that it can be unrolled like the turning of a wheel. This means establishing a base edge which, when the section is in position, guarantees that it will go down square all round.

Coat the deck and the underside of each section with the adhesive, dealing with the foredeck, then the sidedecks, cabin top and so on separately. If the whole deck is coated there will be nowhere to stand while working. The adhesive can be thinned slightly to allow even spreading. The work is then allowed to dry to a state of touch-tackiness which means that a light touch with the finger tips feels quite dry, but very slightly tacky as the fingers are lifted. The sheeting must on no account be picked up while it is wet because it will, at that stage, distort by its own weight.

Usually the best place to start laying is from the centreline working outboard. First lay a narrow strip of polythene to cover the centreline; this will prevent the edge from sticking and aid the butting of the second piece of sheeting to come later. Lower the sheet into place with great care and work slowly to ensure that no air is trapped. The companion piece can now be laid, again starting from the polythene covered centreline. It can either be butted straight up against the first edge or both pieces lapped over it by half an inch.

If it has been lapped (and this is a neater method) a straight-edge can now be laid up the centreline and both thicknesses of material cut with a sharp trimming knife held exactly vertical (Fig 5). In either case a strip of neoprene is now prepared with adhesive on its underside and, with the two edges of sheet pulled back, the

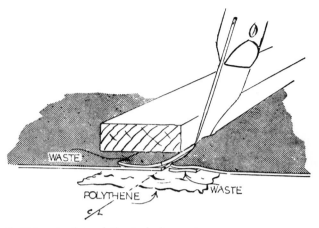

Fig 5. Trimming lapped plastic deck covering.

polythene strip is removed and replaced by the neoprene. Adhesive is applied to its upper side and allowed to become touch tacky; finally, the two edges of sheet can be smoothed down in a butt. The neoprene backing will make it quite watertight.

More neoprene is used in the same manner where the double curvature of the deck edge calls for the sheeting to be nicked to get rid of surplus, and in fact anywhere where there is a gap or a butt. The sheeting must lie naturally. Once you begin heaving it around to make it fit it will be distorted – if indeed it moves at all. Double curvatures call for a slit in the material to accommodate the surplus bulge; don't nail it down. Edges are more difficult than canvas and must be held down by beadings.

Whether working with canvas or plastic these beadings must be fitted snugly and bedded with luting or mastic compound. The ¼-round mouldings which can be bought are all of 90 degree angle, whereas the angles between deck and coachroof coamings, etc are seldom 90 degrees. This means that the

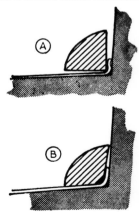

Fig 6. Moulding must be planed to fit the angle.

angle on the moulding must first be planed to fit and the point of the angle planed off so that the moulding can bed right in (Fig 6). It is also a good thing to plane the angle so that its upper edge will fit hard against the coaming, leaving a narrow gap to be taken up by the luting. It is the edges of deck coverings that are its weakness. For further protection of butted edges it is no bad plan to screw down strips of mahogany batten so that they cannot be scuffed up.

5

Glassfibre hulls – repair and maintenance

With glassfibre hulls look for star-like stress crazing around fittings, in way of strength points such as bulkheads and the like. If a fitting has been added and tightened down unevenly, the gelcoat may have cracked and allowed moisture to enter the mat within. A hard blow may have sprung the glassfibre to such an extent that the gel has cracked with similar results, possibly delamination. The main hull will have a number of scratches to be filled, but look out for the really deep ones which have penetrated to the mat. Below decks examine the bonding of wooden bulkheads to shell for signs of cracking away due to excess movement. Faults of this sort should be brought to the attention of the builders. Most damage is likely to be in the nature of beadings and mouldings which, fastened with self-tapping screws, have come adrift; but in cases of bad condensation below, moisture can easily start up rot in plywood lockers and cabin soles. Look at chainplates and stem fittings. If bonded in they should show no signs of cracking in the surrounding GRP. Glass around deck fittings should be checked carefully for the same reason. With hulls that have an inner moulding to line out the cabin any leakage may go undetected for a while. In general, examine the whole hull with care, sighting along the outside for telltale irregularities in the reflected light-shine. There will be small ones; the extent of them depends on the accuracy of the original mould, but major flat areas, humps and hollows should be followed up. Inside the hull, bulkheads, bunks, cockpit fittings, ply panels and so on which have been bonded into place should show no sign of cracking away. Likewise, the bond between hull and deck should be intact and leakproof. If anything serious is discovered get in touch with the builders if she is a first-year boat or have her surveyed by a GRP surveyor if she is not.

Don't be deluded into thinking that GRP is a miracle material. It can deteriorate quite quickly if neglected or misused. Osmosis (blistering) is another defect that has appeared with time and about which not much is known as yet.

Moulded keels are trouble spots. When a hollow trough is moulded as a container for the ballast weights, the lay-up is seldom done properly on the inside and, as a consequence, the glass mat is porous with only the very thin outer gelcoat preventing water from soaking into it. This is the case even with a well laid-up hull but to a lesser degree, and with a bad hull the porosity may be extreme.

The bonding over of the keel on the inside of the boat is another weakness. Badly done it allows water to seep into the keel cavity where it softens and eventually delaminates the mat. By causing the ballast to rust and expand it can also rupture the keel from within. Unless the ballast has been set in a resin mix of some kind, the keel should be regarded a bit askance. On the other hand, it may consist of a one-piece casting either contained in a GRP box or bolted directly to the hull stub. Either of these methods precludes much risk of deterioration.

GRP topsides

Our first impression of a boat is the state of her topsides. She may be well maintained and cared for in every other respect, her spars and sails, engine, electrics and woodwork in good repair, but topsides that are faded and streaky and marred by scuffs and scratches give an impression of overall neglect; this can knock hundreds of pounds off the secondhand market value of a boat.

A wooden boat, whatever her age, can be given an annual spring rebirth. Scratches and dents can be filled, and with average skill an owner can paint a hull and restore it to a state of gleaming beauty. Thus wooden boats, although expensive in terms of time and paint, are highly rewarding in terms of pride and satisfaction. Moreover, if you get fed up with the colour you painted last year, you can change it the next time you fit her out.

The colour gelcoat of a glassfibre hull is very thin and brittle and this makes it vulnerable to damage – small wonder that owners can become as tetchy about their topsides as any house-proud housewife defending her polished dining table. A new boat consequently may be an almost painful liability; even rubber fenders can pick up grit and mar the glossy topsides, dinghies will come alongside to a very half-hearted welcome, and woe betide the novice crew who allows the anchor to thud against the bows. As time passes, though, the colour gelcoat fades, scratches and scuffing, stains and bruises proliferate, and no amount of rubbing and scrubbing with costly miracle hull cleaning nostrums ever seems to restore the original shiny brightness. This is no excuse for allowing a hull to deteriorate.

When buying a new boat the question of choosing a colour should be given careful thought. Any full-bodied colour, especially dark reds and blues, will certainly fade badly in time. Polishing restores the colour for a while, but the topsides are, as the poet put it so succinctly, 'like washen stones, gay for an hour then dulling as the wave goes home ....' White topsides cannot fade but they can become yellowish, and the waterline staining that is a consequence of our polluted seas is more obvious that might be the case on a coloured hull. Pastel shades are a reasonable compromise; they stain as badly but at least the effects of fading are less obvious.

The advantage with a white gelcoat is that prior to the yellowing of age it is easy to get a colour match when making repairs. Repair gelcoat can be bought and mixed in an endless variety of colours and shades, but it takes an expert to achieve a mix that will be a good match when dry. When buying a new boat the builders will supply on request small quantities of touch-up gelcoat of the appropriate shade. This will tide an owner over the first couple of seasons.

With an ageing glassfibre boat that is faded and scarred you have the final option of painting her. Logically it is a sensible solution, although there persists a general feeling that it somehow devalues a boat, and it is a form of capitulation, and that the boat is no longer maintenance-free. However, *no* boat is ever that. Painting is an entirely practical solution and a treatment that both beautifies and protects an ageing hull must surely be good. Neither does painting condemn an owner to an annual chore, because a GRP hull skilfully painted can be kept looking smart with the application of boat wax like any other glassfibre hull.

There is one proviso where painting is concerned. If a professional standard of finish cannot be achieved by an amateur owner, the work is best left to the experts whatever the cost. An old badly painted hull marred by dribbles and curtaining plummets in value, whereas a high-gloss hull looking like new, albeit painted, is a very different proposition.

Routine cleaning

Fitting out really begins with laying up; if a boat is hosed down with fresh water and topsides washed with hot soapy water, then degreased and heavily polished, she is much easier to clean in the springtime. The polish should be a non-silicon boat wax thickly applied and not buffed up to a gloss. In the spring this wax and the grime adhering to it can be removed and the real clean-up can begin.

It is preferable to carry out the main hull-clean in the autumn in order to lessen the work during fitting out. Heavy waterline staining can be tackled first with paraffin. Hull stain removers can be bought in great variety but, effective though they may be on other stains, the waterline and the area just above it are notoriously hard to clean. There are jellified chemicals that can be effective in some instances, although the cold conditions so often prevailing in the spring render them less effective.

A simple but often successful treatment for waterline and rust stains is a 10 per cent solution of oxalic acid (always wear goggles and gloves) bought cheaply at any chemist in powder form. The method is to work away at one small patch at a time, constantly remoistening it, and finally washing off with copious amounts of fresh water. I have yet to hear of a gelcoat being damaged by oxalic acid, but no doubt the generous washing off is a safeguard not to be neglected or skimped.

Perhaps the household bathroom cleaners and car bodywork revivers that call for a good deal of elbow grease are the most effective and the least costly. It must be remembered though that some of these are fine abrasives which reduce the thickness of the gelcoat by a small degree each time they are used. Once a year in moderation and to tackle really stubborn stains an abrasive may be used, but certainly not as a regular treatment.

On deck

The non-slip pattern moulded into decks is a prime collector of grime, which only hot water and detergent applied with a short bristled scrubbing brush (an old nailbrush) or even a suede brush will shift. Deck structures, hatch surrounds, coachroof and coamings quickly lose their original gloss, although regular wax polishing of smooth areas not likely to be walked on will help to preserve it. Once decks begin to look shabby, the painting of non-slip areas with a proper rough deck paint or the laying of adhesive non-slip sheeting (far more expensive) makes a spectacular improvement. Choose darker rather than lighter colours for deck paint as these will be less likely to dazzle crew in bright sunlight – a prime cause of headaches when at sea.

Teak trim

The vogue for leaving teak trim as bare wood and allowing it to weather naturally to a silvery grey is all very well for those who keep their boats in places free from air pollution. A base that lies downwind of cities and industrial areas is a very different matter: grime impregnates and blackens bare teak unless it is scrubbed regularly and frequently. This also applies to yachts with teak-laid decks; only constant scrubbing can keep them looking good.

There are a number of teak cleaners and brighteners on sale which are applied and rinsed off straight away. Stubborn stains can also be tackled with a somewhat risky mixture of household bleach and cleaning powder – risky because mixing them produces toxic fumes only tolerable when used in the open air; the mixture should be applied and then rinsed away immediately with masses of water.

The sheer effort of keeping bare teak looking attractive is too much for most people and all too soon it looks terrible, blackened, and even mossy in hard-to-reach corners. In my view, it is better to oil it or varnish it (see later). Ordinary varnish doesn't take well on oily teak, but a word in passing on the use of teak oil. Never apply it on a windy day and always have a rag soaked in white spirits ready to hand for dealing with accidental drips. Never stow oily rags in enclosed and airless lockers; there have been cases of spontaneous combustion occurring, and smouldering rags can quickly burst into flames.

Osmosis – the dreaded boat pox

Nowadays, owners of GRP boats are all too well aware of the fact that the material is by no means indestructible; and, in addition to the fading of colours and the scarring, scratching and chipping of the hull above water, there is the possibility, indeed near certainty, of osmotic blistering underwater. It can occur at any stage of a boat's life and varies in severity from tiny blisters to deep extensive pustules which penetrate deep into the glass lay-up and constitute a serious weakness if not attended too. Not surprisingly, this osmosis is regarded with the horror and dismay once reserved for the Black Death, although it is not contagious of course. Once contracted, though, it can bring the market value of the boat tumbling down disastrously.

A more realistic attitude is to regard osmosis in the same light that fungal decay is regarded by owners of wooden boats: the damage is irreversible, but it is repairable and to a large extent preventable. Rot in a wooden boat is contagious (unlike osmosis), but in both cases early treatment and the removal of the cause is the real essential.

With GRP hulls the use of a clear unpigmented gelcoat by builders should become standard procedure, for not only can the quality of the lay-up and the presence or otherwise of air bubbles be seen but the absence of colour pigments leaves the gelcoat less water permeable. Buyers should at least be given a choice of clear or colour-impregnated bottom gelcoat. Buyers can often, though, pay a bit extra to have a new boat bottom-coated with an epoxy paint which forms a very good hull protective; and since this epoxy resin can be applied to a totally dry hull and under controlled factory conditions, it has every chance of success.

A hull can be epoxy resin painted by an owner at a later date and there are paint systems marketed specially for amateur application; but here is a warning: no matter how efficient the coating may be, its success depends almost entirely upon the dryness of the hull when it is applied. With an ordinary painting job the exterior surface of a hull can be air dried by a spell of fine weather or by keeping it under covers for a couple of months. In the case of epoxy resins this is simply not good enough. The hull may be surface dry but still water impregnated after a season afloat. Only prolonged drying out under cover, the use of de-humidifiers, and a moisture meter in expert hands can guarantee that epoxy resin painting will be successful in protecting an otherwise unblemished hull.

Having said all this, owners should not become hypochondriacal about things. They should certainly be alert, though, and if signs of blistering occur in a boat that is only in her second or third season the builders should be tackled at once. Hulls that are hauled out every autumn and laid up on dry land are far less likely to suffer osmosis in their first ten years of life than boats that are wintered afloat in a marina berth – particularly if there is a high percentage of fresh water mixed with the salt as might be the case in a river mouth and during the wetter winter months.

The first minor blistering is considered by many people to be unimportant and to be dealt with by popping, drying, filling with epoxy putty and painting over. With annual drying of the hull it may in fact not get much worse, although it is always wise to have a surveyor's expert opinion. This presupposes that these tiny blisters have not yet penetrated deeply into the glass mat lay-up beneath. Once this happens and water begins to spread through the fibres, osmosis proper is about to take place and at this stage an owner must be prepared to take instant and expensive action. Just as a wooden boat with decay spreading throughout her internal construction is on the verge of becoming irreparable unless something is done, so also must a GRP hull be caught in time.

The whole process of drying out, stripping or grinding off the gelcoat and recoating is now commonplace in the marine industry; however, it is not a process for amateurs to tackle although some may have done so. Unless there is access to a suitable shed, and a sandblasting expert can be employed *who has learned how to remove a gelcoat*, the work is far best left to a good yacht yard where sand or slurry blasting or a gelcoat peeling tool may be used. The bared laminate will be dried out under control, voids will be filled, and an epoxy skin built up to 400 microns in thickness will replace the original gelcoat. All of this will cost a great deal of money – an investment to protect an investment.

The first job upon hauling out for the winter and after scrubbing off the bottom is to search for blisters. Use an electric torch shone flat along the surface to detect any pimples. Should any be found, they should be pricked and sniffed at to detect the telltale 'peardrops' odour of the osmotic blister.

Working with glassfibre

The atmospheric conditions, particularly temperature, are closely controlled in boat building factories, and the curing times of the catalysed resins are calculated. Working out of doors or in an unheated shed the case is very different; resins taking longer (sometimes much longer) to cure in cold weather but very much quicker than anticipated in warm weather (so much quicker on some occasions that working becomes impossible). While the proportion of catalyst to resin and the use of accelerators can regulate the speed at which a resin will cure, it is largely a matter for experts – amateurs are better advised to stick closely to the instructions printed on the materials they buy.

Once it has been catalysed, polyester resin will cure faster while in the mixing pot than when it is in use; heat builds up during the curing process which is why it is unwise to add extra catalyst at random in order to speed up the job – cases of the mixture actually catching alight are not uncommon. The usual mixture is a 1 per cent addition of catalyst for a curing time of 30–45 minutes in a temperature of 75° F, and while curing time can be hastened by doubling up the quantity of the catalyst (remember that the pot life becomes very short) this is only permissible in small quantities.

Heat hastens curing time. For every 15° F above 75° F the curing time is *halved*. External heat from hairdryers and electric fires must never be allowed to raise the temperature of the work above 100° F. Alternatively, and perhaps in cool weather, placing the pot of catalysed resin in a bowl of hot water will hasten curing; however, this should only be done if the resin is to be used quickly as it drastically reduces pot life.

When curing, resin goes through a number of stages and a mixture that has a pot life of, say, half an hour starts to gel very soon after that, becoming cheesy about half an hour later (which is the best time to trim off the excess with a *sharp* knife), and rubbery about an hour after catalysing. Final hardening can take as much as two weeks, although the job can be sanded and put into use soon after initial curing.

Fine scratches

These tend to look worse than they are as a result of the dirt they harbour. Clean off any old wax polish with white spirit. If the scratches are very fine they can be softened out by careful application of acetone with a camel hair paintbrush. Acetone is a solvent of glassfibre, so use great caution and neutralise with white spirit immediately. Do not keep a supply of acetone aboard the boat for fear of accident. Finish by using a buffing paste and wax polish.

Deeper but not serious scratches

A gritty fender or contact with a boat pier may cause minor scratches and scruff marks. Use a 600 grade abrasive paper and, trying one small area first, concentrate on sanding down the scratches without penetrating the gelcoat. Finish with a rubbing paste, polish and buff. The important thing is to decide just how deep a scratch may be and whether it has penetrated the gelcoat to the lay-up beneath.

Deeper scratches

In this case the gelcoat has been penetrated. First, using a very fine paper, sand down the area of the scratch and its immediate surrounds, then dust it clean but don't touch with bare fingers. Take a small amount of matching gelcoat and catalyse it according to the printed instructions for the temperature of the day and apply it with a matchstick. Take a small square of plastic film (not Clingfilm) and tape one edge just ahead or above the filled scratch. With a flat-edged scraper drawn across the film, flatten it down on the work and, maintaining tension, tape it in place. When the resin has cured sufficiently, peal off the plastic and with a cork block and 600 grade paper of wet-and-dry type, rub flat – being careful to remove any remaining feather edge. Finally, buff, wax and polish.

Blisters

Blisters (other than the osmotic variety) are often a sign of faulty workmanship; they are caused by small air bubbles trapped in the gelcoat and are often only revealed when the gelcoat caves in above them. Some are miniature craters, others a peppering of tiny holes. Tapping with the handle of a knife may reveal others if suspected. Carve away sharp edges with a fine craft knife, and sand the surrounding area with 600 grade to provide a keying surface. Apply catalysed polyester putty to larger cavities and sand smooth when cured using a flat cork block. Finish with a coat of gelcoat followed by fine sanding, buffing paste and polishing. Lastly, apply wax.

Crazing

A GRP laminate is flexible but gelcoat is brittle, hence a thin laminate or one that lacks reinforcement behind or beneath it is liable to bend under pressure, causing the gelcoat to craze in a mass of tiny cracks. An over-thick gelcoat can produce the same effect. Crazing is often found in the area surrounding a bolted fitting such as a deck stanchion and, while the fitting and its attachment may be plenty strong enough and slight movement is acceptable, the crazing is unsightly. A temporary repair can be made by using acetone or by filling with gelcoat, but in a matter of months this crazing or star cracking will reappear. A more permanent, although not guaranteed, remedy is to sand right down to the laminate and build up a new gelcoat of several layers by brush application followed by sanding flat and re-polishing.

Heavy gouging

Provided no structural damage to the laminate has occurred, these gouges can be dealt with by an amateur. Decide how the damage happened in the first place. If perhaps the gouge was inflicted while

GLASSFIBRE HULLS – REPAIR AND MAINTENANCE

the yacht lay unattended on her mooring, the length of the gouge and attendant longitudinal scratches would indicate a glancing blow from some other boat momentarily out of control. The height of the damage on the topsides gives some indication of the size of craft involved; also, check your rigging and crosstrees as these may have been damaged. In an ideal world the perpetrator will get in touch and confess all, but unhappily there is no guarantee of this. If the damage is short, deep and lower on the hull, it may indicate a head-on smack from a rampaging dinghy. Beware, this sort of impact damage is the worst that can be inflicted on a GRP hull; seek professional advice at once.

In any such damage what must be considered is not whether you are capable of making a repair, but whether it will be of a professional standard. An amateurish repair, however serviceable, will lower the value of the boat in the eyes of the future surveyor. If the damage looks as though it might be more deep-seated than meets the eye, it is a wise move to consider making an insurance claim straight away.

However, assuming that we are merely dealing with a deep but plain gouge, proceed as follows:

1 Bevel the edges and round out the ends of the gouge to avoid a narrow tail to the repair.
2 Fine sand the gouge and its surrounds. Dust and wipe with white spirits to remove fingerprints.
3 If the gouge is deep enough, part-fill with resin putty leaving room for the gelcoat later.
4 Apply strips of adhesive tape parallel to the gouge and a little way back from it, also projecting beyond the ends of the gouge. The thickness of this tape will dictate the thickness of the gelcoat to be applied.
5 Apply catalysed gelcoat with a pallet knife, spreading it level with the thickness of the masking tape and leaving an extra little 'wave' of it at one end of the job.
6 Take a small strip of plastic film and tape one end to the job just beyond that extra thickness of gelcoat. The reason for this extra wave will now become clear (Fig 7).
7 Holding the plastic sheet just clear of the wet gelcoat, draw a flat scraper along the top of the plastic – thus laying it flat and pressing out trapped air in one movement as the 'wave' is spread along the gouge. Repeat the movement with the scraper if any air seems to be trapped.
8 Hold plastic sheet taut, tape down its free end and both edges.

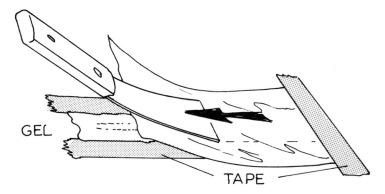

Fig 7. Spreading gelcoat under plastic film.

(Note: It is a good idea to have a practice run on an old bit of scarred plastic laminate before working on the real thing.)

If the gouge is wide and deep, the resin putty should be replaced by a mixture of finely chopped glass (⅛th inch or 3 mm) which has been rubbed between the hands (wear rubber gloves of course) and then mixed with the catalysed gel.

Once the work has cured, hard sand down the repair which will have been left standing proud by the thickness of the adhesive tape. Complete with wet/dry grade 600 paper to remove any feather edge, wash clean and, with a lambswool bonnet on an electric sander and polishing compound, buff up to a final finish. Make sure though that the sander doesn't exceed 2500 rpm or the frictional heat may do damage. Wax and polish.

Repairing a hole

Bearing in mind the folly of attempting a repair that may have unseen but more serious complications, and the obvious need to do a fully professional job, the sort of impact damage likely to produce an actual hole in a hull is really a case for a survey and an insurance claim. Having said that, though, there may be cases in which a hole left by an earlier fitting – in, say, a bulkhead or a hole in a knockabout dinghy tender – may have to be filled.

If the hole was caused by damage, first cut back beyond the extent of radiating cracks or delamination. An electric jigsaw with a metal-cutting blade may be used or a fine-toothed hand padsaw, or a series of small linked holes around the periphery may be drilled with an

GLASSFIBRE HULLS – REPAIR AND MAINTENANCE

electric drill, rasping the jagged edges smooth afterwards. Give the hole well-rounded ends, never corners, then proceed as follows:

1 With a file, bevel the edges of the hole on the outside of the job.
2 Use an electric sander to grind away the surface for a couple of inches all round the hole on the *inside* of the work leaving a clean area. Grind the edges of the hole to a feather edge.
3 Prepare a backing plate by taking a piece of stiff (not corrugated) cardboard and wrapping it in plastic sheet. This backing will go on the outside of the hole; the repair will be done from the inside.
4 Make sure that the plastic film is smooth and tight; tape it on the back of the card to make it so. Place it over the hole on the outside of the hull and tape it down securely. It must lie flush and firm over the hole (Fig 8).

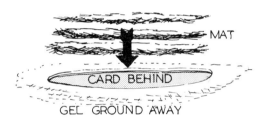

Fig 8. Repairing a hole.

5 Cut a number of pieces of fibreglass mat which exactly fit the hole, either by laying and marking each one separately or by first making a paper template and cutting four at a time from folded mat. It is hard to judge how many you will need, but usually it's more than you'd expect.
6 First brush-apply clear catalysed resin to the hole and backing, then add successive layers of mat – thoroughly wetting out each with resin by dabbing to remove air bubbles.
7 With the hole filled flush, take some larger pieces of mat (at least 2 inches larger all round than the hole) and overlap on to the prepared surface surrounding the hole. Build up this overlapping area to a thickness equivalent to the actual hull thickness.
8 Once all is cured, grind the outside surface level.
9 Apply a second laminate thoroughly wetted out and standing proud of the work surface.

10 Stretch plastic film over the work as described earlier, drawing a stiff edge of cardboard over it to squeeze out any air bubbles; stretch the film and tape it down.

11 After curing, sand down smooth and flush and then apply two separate gelcoats. Progressively, fine sanding and buffing completes the job. Remember, though, strong it may be, but invisible it may very well *not* be.

Painting a glassfibre hull

Since the boat to be painted will almost certainly be elderly or badly faded at least, the likelihood that any of the hull-mould parting agent used during construction being still present is quite remote; however, if it was it would certainly ruin a paint job. So also would the presence of boat wax containing silicones. Carry out a check by pouring a little water over the topsides. It should lie or drain down in broad sheets; if it breaks up into globules, the hull must be carefully degreased. In any case, the topsides must be thoroughly washed, dried and wiped over with white spirit and *not touched by human hand thereafter.*

It should be mentioned that having decided to paint, one should settle for one paint manufacturer and stick to the system and products advised by that company; if possible, assemble your materials before you begin, ordering more than you need on a sale-or-return basis – most chandlers will oblige.

From here on much depends on the individual manufacturer's advised procedure. One may recommend giving the whole surface to be painted an even rubbing down with a fine grit paper, thus removing any trace of grease and producing a good keying surface; others offer a special primer which has the effect of slightly surface-softening the gelcoat to give better paint adhesion. This primer will have to take its first over-painting coat within twenty-four hours.

And here a comment on outdoor painting. Ideally the whole job should be done under cover and in dust-free conditions when air temperatures are average and there is no risk of frost. If this is not possible, it is then very important to study the weather closely and choose a period when barometric pressure is stable or high, giving the almost guaranteed likelihood of at least three dry calm days. Of course in true anticyclonic weather this also means a strong possibility of foggy or misty mornings, and the whole boat will be damp even if previously covered over. Do not paint if it is foggy; and even though it shortens the working day drastically, do not start painting

GLASSFIBRE HULLS – REPAIR AND MAINTENANCE

until the whole hull, including decks, are bone dry. The materials you will be applying are highly vulnerable to moisture.

The enamel will be either a one-pot or a two-pot system, the latter – although less convenient – having better weather resistance in the long run. With the two-pot enamel the two components (base and hardener) start to cure from the moment they are mixed, having a pot life of four to five hours and a drying time after application of two to three hours depending upon air temperature. After mixing, which must be done gently in order to avoid whipping up froth, the enamel must be allowed to stand for perhaps ten minutes to allow any traces of air bubbles to disappear. Be very careful over this stage; *under*-mixing can mean that the enamel will never cure properly and remain sticky for ever more, while vigorous *over*-mixing will result in a paint finish ruined by pin-holing.

Technique

A second coat can be applied six hours after the first coat, although second-day application is usually better and more convenient; never

Sacrificial anode. Placed on the outside of a hull and wired internally to potential sources of electrolytic corrosion, the anode wastes away instead of propeller, shaft or other fitting. Note the deep pitting – evidence that quite serious electrolysis is present and indicating a need to identify it and to replace the anode.

rush to slap a second coat on unless you have made a serious miscalculation of the weather prospects and you are painting outdoors. If the interval between coats is more than a couple of days, a fine grit rubbing down will be needed to provide a keying surface.

The application of paint to large smooth surfaces is the most taxing test of a painter's skill and, without reiterating my strictures on the need to achieve a professional finish, it is no bad thing to paint the glassfibre dinghy tender first to get the knack and the self-confidence. Each paint needs different handling according to its thickness, spreading qualities and rate of drying, and while the professional painter sums up its nature at a glance the amateur may be well on the way to ruining the job by the time he gets the feel of it. If the yacht has a transom stern, it is quite a good plan to start there – where errors are less obvious.

Depending on atmospheric conditions, sun, and drying wind, paint is either easy to work, smooth and flowing, or it is the very opposite. Never attempt to paint if the work is exposed to hot sunshine. Even on a warm day – despite being sheltered from direct sun – paint can dry at such a speed that every brush mark remains visible. However, if the work *has* to be done, follow maker's instructions carefully and 'let down' the paint with the recommended thinners. Remember, though, that while this will enable the paint to flow, if it is too thin by even a very small degree it will be prone to drips and dribbles or 'curtaining'. It is when you return to try and paint out these dribbles that the surface is marred the worst.

When applying ordinary paint as opposed to this two-part polyurethane, the usual technique calls for laying on the paint with horizontal then vertical strokes; you should then use oblique strokes before finally laying off with level horizontal ones again. This guarantees an even distribution of paint. The polyurethane paint cures too quickly for this treatment, so for this we use a 'wet edge' technique. The aim when applying this first coat is to get as much paint on to the surface as possible, but consistent with it lying smoothly and not hanging in curtains. Accordingly, it is applied in vertical strips of 3 or 4 inches in width and at a pace that allows you to keep the wet edge of the paint blending with the following strip as it is applied. Keeping the wet edge moving is what it is all about; once this wet edge is allowed to dry, even slight brush marks will remain when you try to blend in the next strip.

It can be seen then that the area of fresh paint you can handle while keeping the wet edge moving (looking back to catch any drips that may be starting to dribble down) is entirely dependent on the rate at which it is drying. A professional painter once told me that you should do nothing until you had come to terms with your brush. Use the biggest that you can handle and make sure that it isn't going to lose hairs, never dip it deeper than a third of its bristle length into the paint, and never, never *dab*. The ideal then is smoothly flowing paint, even strokes with a generous brush (2 inches or bigger), and a drying rate that allows time for you to stand back now and again and take a good look at the overall job.

Once this first coat has dried it is time to use a trowel cement to level out any visible irregularities. Leave it slightly proud, and then when it has set hard grind it down flush. If much filling is done, finish by giving the whole area a fine grit sanding followed by a wipe with the appropriate thinner. A total of four coats may be needed, fine-sanding between each, and scrupulously removing every trace of dust between each. The final coat may be applied by lambswool roller, bearing in mind that it may be cheaper to buy a new roller than to clean it with expensive thinners. Some professionals achieve their highest gloss by using the old coach builder's technique of mixing enamel with a compatible polyurethane varnish in equal parts. Waiting a month for the surface to become fully cured and then buffing with an electric buffer and paste, and finishing with a wax polish, is equally effective.

6

Painting wooden craft

It is the paint – the colour and shine – of a small vessel that singles her out as a yacht irrespective of her shape and type. Many a fishing boat or yacht built as a fishing boat would pass as a working craft were it not for her finish. In some ways it is a convention that has become a heavy burden upon us. The tarry hulls of old time craft have a lot to commend them. Yachts are yachts though, and a shabby vessel of any kind is an indictment against her owner.

The real purpose of paint is protection; however, it won't protect if it is badly applied and the basis of all painting is good and thorough preparation. Of time spent on the job, at least two-thirds of it should be preparation of the surfaces. There is a difference between bright paint and gloss. Any paint job will look attractive for a week or two, but thereafter the rough surfaces will become grimy and any shine will vanish.

When all the minor repair work has been completed, the type of painting must be decided upon. This can vary from a rub down and a couple of new coats on a sound foundation of last year's paint to a complete stripping of old paint and a build up from scratch. A re-paint such as this may only be needed every five or six seasons if the original work was good and good-quality paint was used. On the other hand, it may be needed every third year or so.

Burning off

Stripping would be a fairer term since total removal may also be done by using chemical stripper. This is slower and much more expensive, but there is less risk of damage to the surface. Do not stint the paint remover. Apply, wait for the paint to soften, scrape without using force, apply more to the same patch, and so on until bare wood is reached. Many people combine paint remover with vigorous use of a sharp scraper and in so doing cut up the wood. Use a sharp scraper, but use it gently and let the stripping fluid do the work.

Some strippers must be neutralised afterwards with white spirit, others by a wash with fresh water; be sure to do one or the other. Be sure also to prevent the stripper from splashing on Perspex or other plastic windows as it will 'burn' deeply into them. Dry scraping with a very sharp Skarsten scraper, with a hooked edge blade, is an alternative for paint that is old and powdery. Never be tempted into scraping across the wood grain. Although this gets the paint off more easily, the scars that remain will take a lot of sanding out. With care a hull may be dry scraped to shift all the loose and easily moved paint and then stripper used on the remainder.

Burning off with a blowlamp is by far the fastest and most efficient method. If the lamp has to be bought, start off right away with a bottled gas lamp – this is provided the yacht has gas bottles already in use, otherwise the whole procedure becomes expensive. The method is as follows.

Choose a day without much wind if working outdoors otherwise most of the heat is lost. Check that there is nothing lying around under the hull that is obviously inflammable and, to be quite sure of safety, have one or two old wet sacks and a bucket of water handy. For a right-handed man it is better to work from right to left so that with the lamp in the left hand and the scraper in the other, the paint is being softened in advance of the scraper as you move along. Play the flame on the surface until the paint lifts and heaves in bubbles, flip the flame aside and scrape, play the flame again, flip aside, scrape and so on. Do not heat the paint until it blackens and catches alight. This will happen from time to time and is not serious, although the flaming strips of paint that fall should be stamped out. This overheating does however char the wood and bake the seam stopping so that it becomes brittle; the charred wood when scraped naturally leaves a scar. With practice, the hull can be stripped with very few singe marks to be seen.

Use a flat painter's scraper and lift the softened paint. The topcoats will come off easily while the undercoat may become sticky and hard to shift. Let it cool and it will be brittle and easy to clean with a sharp Skarsten. It is quite a good plan to burn off only the easily lifted topcoat and follow round later with chemical paint stripper cleaning off the rest.

Try to avoid leaving a bare hull out of doors overnight. A coat of metallic bare wood primer should be applied or the hull well covered over with sheets to keep out the damp. The hull, being bone dry from the blowlamp, will absorb moisture quickly enough from the air in any case.

After burning off, get to work with an electric sander using a coarse number 4 aluminium oxide or grit. Do not use a rotary disc sander though or the curved surfaces will be badly scarred. Use either an orbital type or hand sanding on a cork block, always with the grain. Having dusted down, the bare wood primer can be applied if it is time to pack up for the day, otherwise proceed with the sanding, continuing with a grade 50–80 grit and finishing with a 120 grade paper. Use only the very best make of paper. The wood surface should now be satiny smooth save for the scratches and indentations – the hills should have been cut down leaving the valleys.

Now apply the metallic bare wood primer. If this was done earlier for the sake of protection, the subsequent rubbing would by now have removed most of it again except for the hollows and holes. An alternative for the perfectionist, and one that has a lot to commend it, is to use a wood grain filler. This is a dark paste which has to be rubbed on with a piece of moist hessian cloth so that it fills every minute crevice in the grain. When dry it can be sanded flush; a primer coat can then be applied.

This first coat will show at once where the uneven areas are and we must digress a little. A 'filler' is a compound for filling the grain and small scratches while a 'stopper' is a heavier-bodied compound for filling larger holes such as seams and butts and deeper gashes and indentations; both should be non-shrink. Transparent fillers are obtainable for using under varnish.

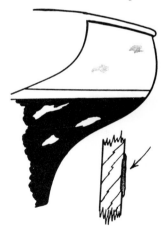

Apply the stopping with a broad bladed painter's trowel (some stoppers are in fact called trowel cements). When it has set, get to work with the sander again and work the whole surface over (Fig 9), flatting down the stopping to the uniform level; wipe over the work with a rag moistened with white spirit and apply the first undercoat. When dry this will again show any hollows still to be treated –

Fig 9. When stopping has set, make sure that the surface is sanded to a smooth, level finish.

and there will be plenty of them. Continue stopping, sanding and undercoating, with dusting between stages, until the hull is even, and satiny smooth. Finish with the finest grade of paper, dust, and the hull is ready for one, or perhaps two, topcoats.

The technique advised by makers of one-can polyurethane paint is quite different, except that the bare wood preparation and stopping is the same in the initial stages. This paint is really a polyurethane varnish with colour added and, like varnish, the work is built up with successive coats of the same stuff. Each coat is fine sanded and dusted. Perhaps four coats will be sufficient but six is more usual. Note here that polyurethane and not epoxide resin paint is involved. The latter is much harder when set, but it is also more difficult to handle. It cannot be applied over old paint and it is highly sensitive to moisture.

Patching up

With a sound paint build-up remaining from previous years, there is no need to strip right off. To assess the state of the old paint, rub a hand over the topsides to see whether any colour comes off. Some paint tends to 'chalk out' when its gloss quality has failed.

Using wet-or-dry paper with water and a cork block (do *not* use a wet paper with an electric sander as it can be highly dangerous), grind down the old paint surface until all trace of gloss has gone. Wash it off and a piebald surface will remain where the high spots, blisters and loosened paint has been ground away. Sand over once more to pick up any areas missed the first time, wash down, and wait until the hull is bone dry.

There should be no sharp edges of paint film showing anywhere, blister patches must be sanded away around the edges. The bare wood must now have primer dabbed over it and, when dry, stopping can be carried out as described. A further rub down, a dusting (or washing if wet/dry paper was used), and the hull is then ready for two undercoats, each with a fine paper rub after it, then a final undercoat and a top enamel finish.

Renovation of really good paint is simpler still. A fine, wet rub down, stopping if needed, an undercoat and a top enamel will probably serve. Small pitting may be filled by a 'scraped on' coat of enamel – use an artist's palette knife as shown in Fig 10. In the case of polyurethane one-can, the rub down can be followed by a couple of coats of the enamel.

Incidentally, the point about undercoats versus top enamel is that the undercoat is denser and has a better colour coverage, while an enamel is more transparent and covers poorly over a mottled hull.

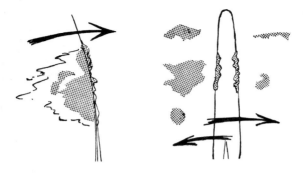

Fig 10. Using a palette knife to fill pitting with enamel.

Painting technique

Buy the best brushes obtainable – say, two 1½ inch, a 2 inch, a 2½ inch and a couple of soft 1½ inch varnishing brushes. Never attempt to clean a brush used for colours to paint white or light shades or vice versa. Have a good stock of white spirit for cleaning them and never leave a paint-loaded brush for more than ten minutes without either cleaning it, suspending it in white spirit, or at least wrapping it tightly in clean lint-free rag to exclude air. After the day's work, suspend, not stand, each brush in its own bottle of boiled linseed oil and wad some paper in the open top to keep out dust. Brushes may be stored like this permanently without ever cleaning them, providing each brush is kept for its own particular colour or for varnish and that the oil level is kept topped up.

Too much cleaning is bad for a brush because some paint always remains to become granulated and thereafter to find its way down to the bristles. If a brush is to be cleaned and stored dry, use a paintbrush cleaning fluid worked well into the head of the bristles, then wash in detergent followed by fresh water and allow to dry. A new brush or one that has been stored dry should be rapped smartly on a table edge to dislodge dust and old paint specks, loose hairs, etc.

Never dip more than half the bristles in the paint and don't dab – instead, use long, even strokes. Undercoat can be worked to fill the surface, but enamel and topcoat should be laid on with first vertical then diagonal and finally horizontal strokes with the grain (Fig 11). The big danger when using enamel lies in applying too much paint and causing runs and curtaining. Once the paint has set off, further brushing will mar the finish; so while painting, keep glancing back at areas just covered to see that runs are not appearing.

PAINTING WOODEN CRAFT

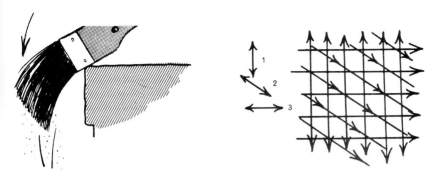

Fig 11. Left: tap brush on table to remove dust. Right: paint should be laid on vertically, then diagonally and lastly horizontally.

With top enamel dust is the real enemy. Use a 'tack rag' or a white spirit moistened cloth of non-fluffy type to clear the surface ahead of you. Don't paint (or varnish) on a breezy day and sprinkle the floor or ground with water before starting work. Watch the weather too. In cold weather out of doors the hull will perhaps have frost on deck which will soon melt on a fine day, but it may be an hour before the first trickle of water runs down the topsides. Dry the decks first and wait until the air has warmed up. Don't paint after about 3 pm on winter days because after that time there may be a dew.

Polyurethane paints can be applied more thickly than others and they also set off faster. Remember that if a final topcoat is disappointing it can still be given a fine rub down and an extra coat. Roller painting for applying undercoats is quite all right and even topcoat can be applied this way, but a brush is still needed for edges.

Antifouling paints

Antifouling paint consists of a mixture of toxic chemicals contained in a paint carrier, which is designed to waste away at a measured rate permitting a steady escape of the poisons. A cheap paint releases its poisons quickly and the coating loses its effectiveness by mid-season. There is a wide variety of antifoulings, using different chemical systems, and it is important to know which are best suited to your particular area or need.

Certain types of paint have been outlawed on environmental grounds, the TBT (tri-butyl-tin) formula was proved to be harmful to marine life and in due course was banned by law. The paint chemists came up with alternatives, however, which are regarded by many users as superior to TBT.

Boats that dry out at low tide may impair certain paints; moreover, in muddy waters a caked film of mud does much to interfere with their working. A common enemy, though, is the heavy copper-based paint, which while being effective may also do much electrolytic harm to underwater boat fittings of incompatible metals.

The usual practice is to coat the hull before springtime launch, but some owners prefer to haul or dry out in late May to coat their hulls, just prior to the hot days when fouling begins to grow at speed. A few antifoulings have limits set on the time they can remain in the air before losing potency, but most are good for up to six weeks prior to launching.

The fouling includes weeds of various types and barnacles, both the small and large barnacle type. Wooden boats have also the risk of gribble worm (Limnoria) and Teredo worm. The former destroys surface wood, moving inwards to sound timber steadily and expensively, but Teredo is the major threat and can be found nowadays in many northern latitudes including Great Britain. In the tropics the Teredo can travel up to 4 feet along the inside of a plank and attain the whole diameter of ¾ inch.

Copper sheathing, the old remedy, is now uneconomic, but the extra strong tropical antifoulings are held to be suitable protection *provided no areas of bare wood* are allowed to escape. Damage during launching, grounding, etc can leave vulnerable bare patches. Glassfibre sheathing is an alternative, but again underwater damage can peel it and leave large areas bared. A newly built hull deeply impregnated with wood preservative prior to painting is a further protection since the Teredo larvae would probably be destroyed before they could grow and penetrate.

There is only one golden rule for the application of antifoulings: *read maker's instructions and follow them.* Usually, the practice is to apply two full coats plus extra thickness at the turn of the bilge, to the keel, deadwood and other favourite fouling areas. The use of antifouling boot-topping is not very effective as the toxic content is lower, sacrificed to paint adhesion qualities, and it is more popular nowadays to carry the bottom paint up and above the loaded waterline.

A warning about shores. When laid up ashore supported by wooden shores and wedges, do not tamper with these in order to get at areas of hull hidden by them; in fact, most yards strictly forbid owners to shift their own shores. In the event that no such embargo exists and you must therefore shift your own, first note the angle of the shore to the hull, see that the shore will be butting down on firm level ground or on a wooden block, and firmly tap in the wedges with a hammer (do not tap heavily as this could buckle the hull and do permanent damage).

Varnishing

The real professional Boat Show gloss is not difficult to achieve. It calls for a combination of careful preparation, complete absence of dust while working, and the requisite number of coats applied with the right brush and the right speed. How long it will last depends upon the quality of the work and the varnish.

Blackened, loose or discoloured varnish must be removed. It cannot be burned off without charring the wood (which will show through later) and so it must be removed either by dry scraping or the use of a chemical stripper. Every scratch and scar left by the scraper will mar the finished work. With paint such scars can be disguised, but it is not so with varnish even with the use of coloured fillers.

Remove the hard old varnish in layers using stripper. Dry scraping is a bit dangerous. Scrape with the grain and study the run of it on curved toerails and the like because a scraper meeting short grained wood immediately cuts it up in a series of ridges (Fig 12). Don't allow an unskilled helper to use

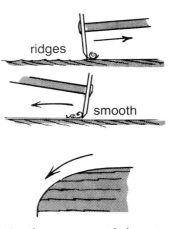

Fig 12. Always scrape with the grain.

a scraper until you have pointed this out. If the woodwork is on a GRP hull, though, keep a sharp watch for chemical stripper splashes; these will damage the gelcoat and, in the case of a hull with a GRP deck, it is better to rely solely upon dry scraping if all the varnish needs removal due to long neglect.

When all varnish is stripped, 'kill' the surfaces according to instructions for the particular stripper used and, when quite dry, begin sanding down, using progressively finer grit papers dry. Wet rubbing on bare wood raises the grain. Dust the work and the entire deck before doing anything else. If the sufaces are still showing scratches and open grain next apply a good transparent grain filler – this is in any case a good plan as it is worth at least two coats of varnish in terms of final gloss; but if you have not had experience of grain filler, experiment a little on some less noticeable part of the work. There may also be need for stopping over screw heads and in deeper scratches, but choose a tint that will be a fair match for the surrounding wood when it has been darkened by varnish.

The first coat is by tradition a thinned down one which can soak well in. This must be sanded down with fine grit, dusted most carefully, and then a succession of coats, each rubbed down and dusted, can be applied. Allow each one to dry completely hard before rubbing and choose a windless day for the work. Avoid working early or late in the day, avoid hot sun or drying winds, and in colder weather stand the varnish tin in warm water for half an hour before starting work. A good plan too is to float a disc of paper on the surface of a half empty tin when leaving it overnight; this will prevent the forming of a skin.

Brushes should be suspended overnight in linseed oil, which must be rinsed out with white spirit before use. Never shake a tin of varnish, and if any skin has formed strain the whole lot through a fine nylon stocking. Try to work so that you are moving away from freshly varnished surfaces.

Choice of varnish is not easy as you may be torn by conflicting advice from friends. In general, the modern polyurethane varnishes dry quickly but do not necessarily set as hard as the non-polyurethane types which take longer to set off. Thus for a quick job, applying several coats on the same day, the quick dryers are excellent. The ultimate test is time though and a hard finish that does not discolour is what we all want. Providing the surfaces were dry and that conditions were satisfactory and the work done properly, a slower drying varnish (6–8 hours) of pale colour probably gives the best results.

The propeller and impellers

It is generally held that bronze propellers should not be antifouled. However, if a propeller cannot be buffed up to a bright finish, I see no real harm in painting it if the boat is to be left idle for long periods allowing weed and barnacles to run riot. Much must depend upon the electrochemical reaction of propeller to paint and, if in doubt, ask the manufacturer – a phone call is cheaper than a damaged propeller.

The impellers of logs plainly must not be painted. Echosounder transducers – again generally held to be best left bare – are none the worse for a thin lick of antifouling; after all, there are many in-hull echosounders that work through the thickness of the hull. Engine and other raw water intakes should be painted, but not to the extent that water entry is restricted.

Teak-oil varnish

The dense grain and oily nature of teak makes it hard to varnish successfully – that is, it seems to shed ordinary varnishes in no time at all. Although teak oil applied regularly is the usual alternative, there is also available a teak-oil varnish which combines the virtues of both treatments. Consisting of a thin gloss and a thicker matt one, the method is to apply to bare wood a thinned coat of gloss followed by three or four further coats of un-thinned gloss and followed by a coat of the matt oil. The final effect is not the high gloss of conventional varnish, but it is none the less clear and hard wearing.

Gold leaf in the form of self-amalgamating tape is applied by stretching it to activate the chemical action which makes it adhesive. In this case, the tape was overstretched and has broken down.

Cove lines

The groove just below deck level on the topsides is often left plain simply because it is difficult to paint in neatly. The best way is to lay on masking tape exactly along the groove, filling it in effect. Having got it true and straight, lay other strips above and below it exactly edge-to-edge. The first strip of tape is then peeled off, leaving an exact width of groove to be painted between the remaining strips. Alternatively, adhesive tape in the desired colour can be used. Choose a windless day to apply it and don't attempt it unless you have a steady hand and a good eye. Traditionally, cove lines are laid on in gold leaf but a modern equivalent is self-amalgamating gold tape. This type of tape is non-adhesive until it is stretched, whereupon a chemical change takes place causing it to adhere either to itself, in the case of a binding, or to a clean dry surface. In practice, and in my own experience, this tape should not be stretched as much as the makers recommend as it tends to shrink and break up leaving gaps.

Cutting in waterlines

Every boat has her designed Load Water Line which is the level at which her designer intends her to float when in trim and upright. This may be a simple line or it may be a broader band painted with a contrasting boot-top paint and extending above the actual LWL. The line itself will have been scribed in on wooden hulls at the time of building and it may be moulded in on GRP hulls. With time, and successive scrapings and paintings on wooden hulls, the original line may have been lost and likewise on a GRP hull only one line may have been provided so that the boot-top will have to be painted in either by eye or by marking it in more scientifically.

In the case of an old wooden hull there is almost always some ghost of the original line remaining and which will be visible when the hull is burnt off, but it will still have to be marked in again in order to be able to follow it with later coats of paint because the initial priming and undercoats will soon obliterate the ghost of it. The easiest method is to use masking tape, picking up the original ghost lines and truing up the rest by eye. This can be pencilled in and then, using a whippy batten held by helpers, scribe in a new line with a bradawl. Beware though as the line of grain will guide the tool away from the line. On a GRP hull the line must not be scribed because this will be damaging to the gelcoat.

With the LWL marked in, it is then possible to mark in the top edge of the boot-top by the spirit level method. The hull must be quite vertical and, with the spirit level fixed to a short batten, either use a ruler set at the height of the boot-top and held at right angles to the batten or use a set square, made up for the purpose. Move the batten along the line and tick off the boot-top height as you go, sliding the square further back as the hull slopes away under the stern. This method can be used in reverse to mark a line below an existing one (Fig 13). Here, the top of the square travels along the existing line and the new one is ticked off below it. Finally use the masking tape to true up the line. When no line of any sort is visible the LWL must be marked in afresh. We must assume that the level at which the boat floats is known, either from the uneven edge of the antifouling or simply from observation. The hull must be vertical and also chocked up so that, seen from the side, she is at the same angle fore and aft as she will be when afloat; failing this, the height of the plank used must be adjusted. The LWL is marked on the stern and on the stem, also amidships. If this is difficult to do, mark her amidships on each side.

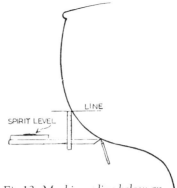

Fig 13. Marking a line below an existing one.

Next set up a horizontal plank at bow and stern as shown (Fig 14).

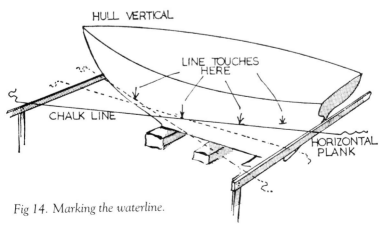

Fig 14. Marking the waterline.

Take a builder's chalk line, load it with chalk by running it over a lump of chalk, and station a helper at one end. Take the other and stretch the line so that it lies taut from the top of each plank and move it in until it touches amidships mark. If the planks are at the correct height the line will now run from plank to plank via your mark. If the taut line is twanged it will deposit a line of chalk around the middle of the hull fore and aft. The line can now be angled round until it touches other parts of the hull (shown by dotted lines) and the twanging method repeated until the whole waterline has been marked in. Run a batten and pencil line round before the chalk is rubbed off and then start work on the opposite side of the hull. Where there is a waterline visible on the side of the hull but lost on the other, a simple way to mark it in is to use the hosepipe method. See that the hull is quite vertical and take a length of transparent hose, long enough to go under the hull and up the topsides to a point above the waterline area on each side. With both ends of the hosepipe held up, top it up with water until the level reaches the visible waterline on one side of the hull; the water level in the hosepipe on the opposite side of the hull will then indicate where the missing waterline should be. Repeat this operation at a number of points along the hull and then use the batten to finish off.

A further point about using masking tape. It can be used as a sight-line for drawing in a waterline and also as a means of getting a clean

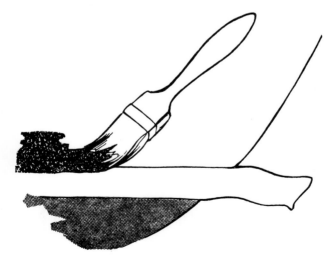

Fig 15. Using masking tape to get a clean edge to a painted line.

edge to a painted line. In the latter instance a final coat of paint will be brought downwards to overlap the tape which has its upper edge along the desired line (Fig 15). Tape is then stuck over the boot-top paint with its lower edge forming the limit of the final coat of bottom paint. When stripped off, a hard crisp line will be left. It is very important though to apply masking tape only after the paint to which it is sticking is not only dry but hard, otherwise when the tape is removed it will lift the paint beneath it. Alternatively, lift the tape as soon as the newly applied paint becomes tacky.

Cutting in by hand is a little art which must be mastered though, as it is not always possible to use tape. The aim is to cut in *upwards* wherever possible so that the runs of paint or varnish will not mar the work below it. The brush must have a clean, thin edge – certainly no whiskers or bushiness – and it must be loaded with enough paint to allow a miniature 'wave' of it to be carried along the line. Don't dab; use firm, level strokes. Begin just below the line with the first stroke then take a second stroke over the same ground but above it and edging up to the line.

A simple formula for estimating paint coverage is as follows:

Topsides:	LOA	(in feet)	+ beam × 2 × average freeboard
Bottom:	LWL	(in feet)	× beam + draught
Decks:	LOA	(in feet)	+ beam × 0.75 – deck structures

The answers are in square feet.

7
Masts and rigging

Recently it has become more common to find yachts laid up with their masts left standing than when they are unstepped and laid out on racks or on deck. However, leaving the mast standing during winter lay-up has one serious obvious disadvantage; it means that neither mast nor standing rigging get the careful examination that is so vital.

If the mast is to be left standing and if it had been highly tuned with standing rigging highly tensioned, this should be eased off all round before the boat is hauled or craned out in order to rest it and because a boat ashore often sets up different stresses to a boat afloat. Running rigging should be unrove leaving messenger lines rove in place ready for re-reeving in the spring. Halyards left up all winter come in for a lot of chafe, and if the yacht is to leeward of a city or an industrial area they will become deeply grimed with acid and sulphurous fumes.

Ideally a mast should be unstepped every year, but certainly every other year or every third year at the very outside. In the event that the mast was unstepped and stored on deck or on racks in the open air it is wise to block up the open lower end of an alloy mast otherwise there is a strong chance that by springtime there will be a busy nest of robins therein and a delayed launching date. The standing wires should have been examined when removed in case any replacements are needed. Take a rag, lightly oiled, and wipe each wire. Any broken wire strands will be quickly detected without damage to your hands.

Stainless wire rigging of the usual 1×19 wire is broadly reckoned to have twelve useful years of life; thereafter it becomes increasingly suspect by reason of metal fatigue. Much depends upon the boat. In the case of a stripped-down racer, standing rigging will be the smallest size of wire consistent to the high-stress use it will get, whereas with a family cruising yacht the rigging may be a shade over

Tailing on a messenger. Removing a rope halyard from a mast and leaving a small recovery line in its place to facilitate replacement of the halyard. Terylene sewing thread is used.

size in the first place and seldom either set up tight or fully loaded in use. Beware of broken ends of wire or kinking. When wire is past its prime, it is a good plan to begin replacing wires each year – starting with the most highly stressed ones.

An alloy mast should be washed with warm detergent suds and white spirit used on stubborn stains. While cleaning, examine carefully for minute hairline cracks around fittings, look for damage to the anodised surface, and check for corrosion. A metal mast is a piece of engineering, whereas a solid wooden mast is inherently as strong as a tree of its size might be. A metal tube standing on end under compression cannot afford to have any weak spots; it is standing in column supported by the various pull of wires – should anything happen to impose unexpected stresses it can buckle and collapse completely.

At the least sign of white powder around a fitting, suspect corrosion and discover its cause. It is quite possible that a fitting has been added that is made of an incompatible metal or that self-tapping screws of unsuitable type have been used and the salt environment is causing electrochemical reaction and wasting corrosion. This *must* be put right, professionally if needs be.

Areas where the anodising has been chafed away, perhaps by an exterior halyard, should be either painted with a metallic spray or just varnished. Never use abrasives on anodised surfaces; if they are stained and cannot be cleaned, just leave them. A fine bathroom cleaner is permissible if used sparingly. Clean the luff groove by pulling a tightly rolled plug of rag soaked in white spirit through it (this can be done when the mast is standing provided that a stout recovery line is attached above and below) and lubricate it with a specially formulated Teflon lubricant. On no account use ordinary oil as it soon metamorphoses into a stiff black paste which inevitably ends up on the sails.

Check over all electrics carefully while the mast is down, especially those at the masthead. A mast can vibrate viciously at times and connections and contacts are under a lot of stress. In the case of a standing mast somebody must go aloft and descend slowly, all the while checking, lubricating, cleaning and examining. Look for bent or worn pins, stiff or non-revolving sheaves, loosened locknuts, missing split pins. (*Note:* Split pins should be opened to 15 degrees and *not* curled right back on themselves.) Ideally the climber will be the lightest person who is experienced enough to tackle the various tasks and pass judgement on the condition of things. In the husband and wife team it is usually the former who goes aloft – how he gets there is another matter to discuss. This will be dealt with in due course; suffice to say now that the mast must be thoroughly checked by *somebody*. The boat should not go to sea until this has been done.

Wooden spars

Wooden spars need careful examination each year to guard against rot and other weaknesses. Old, blackened varnish must be removed not only because a pale, varnished spar is a gladsome thing to see, but because clear varnish is a window through which the state of the spar can be kept under survey. Use a stripper and a sharp Skarsten scraper but be careful to avoid scraping 'flats' on the rounded surface or roughing up the surface by running against the grain.

A pole spar that does not have a sail track attached to it – a gaff rigger's mast for instance – can be bare hand varnished. The advantage of this is that one's sense of feel allows the varnish to be laid on at an even thickness and the dry feel of bare or thin patches is easily detected. Spill a little varnish into the palm of one hand, apply it to the spar, and then proceed as if greasing it down. Surprisingly, very little is spilt – less than if using a brush – and the warmth of the hands keeps it flowing nicely. Hands are more easily cleaned than brushes as well. Spars that have fittings and metal luff tracks fixed to them must be varnished with the brush. At all costs see that no varnish enters the track or luff groove as this will cause the sail to jam when lowering or setting. Follow normal varnishing procedure in either case, rubbing down between coats and so forth. Build up from bare wood to at least five coats.

Longitudinal shakes, which are harmless and due to the drying and shrinking of a grown spar, must be stopped to keep out moisture. There are good spar stoppers sold which are soft, and remain soft, to allow for expansion and contraction, but a reliable old recipe is to use equal proportions by weight of bee's wax and tallow, melted together and forced deep into the shake. If the mixture is too thin when cool, cut back on the tallow a bit. In the case of a built mast consisting of glued sections or perhaps two halves glued together, see that the glue line is intact. If it is gaping in places, consider whether the spar is safe as it may be no remedy just to fill the open joint with stopping if the glue has deteriorated along its whole length.

Solid spars, either 'rickers' (natural trees trimmed to shape), glued spars or spars cut from a log eliminating heartwood, are all prone to rot. A ricker, having sapwood at its centre, is especially prone and such a spar can become infected and rot internally for its entire length without showing traces on the outside. Other types being free from sapwood will usually show signs of decay at mast bands, under cleats and at other places where damp has been trapped. Darkening of the wood, softness, possibly saturation as well, are the indications. All fittings must be removed if rot is suspected.

Wooden spars can also be 'sprung' due to sudden overloading following, perhaps, the parting of a shroud or by (in the case of a boom) a bad gybe. The fracture of the spar may not always be visible at the time, but any wooden spar that has been seen to bend far in excess of what seems reasonable should be suspected. Transverse cracks may appear and, if they do, the spar should be replaced. The longitudinal cracks or shakes seen in ricker spars are not serious being due to the shrinkage of the spar, but the shakes must be soft-stopped to keep out moisture or they *will* become serious.

In the case of a glued-up hollow wooden mast, the spar will have a series of solid sections in way of stress points and main fittings; elsewhere, it is built with a series of voids. There are also simple four-sided box masts in which planks of knot-free timber overlap in scarfed joints, each joint staggered to avoid the creating of weaknesses. There is also the glued-up 'built' solid mast. All are extremely strong, but heavier than a comparable alloy mast by perhaps 50 per cent. In every case it is the integrity of the glued joints that must be watched for – and rot as hitherto mentioned. In some of the old glued masts adhesives may have been used that have since become brittle from age, and unfair stresses have been known to crack these glue lines.

Working aloft

If you have no head for heights, even the modest 30 feet or so of a typical mast will seem terrifying and a nervous person will be hard pressed to get up there – let alone do important work aloft. If you are unhappy up there, there are problems getting up there, or if you don't really know what to look for having got there, then don't hesitate to call in your local rigger. Gambling on the fitness of the mast is like gambling on the accuracy of the compass.

For getting aloft the first priority is safety. You must have total confidence in the state of the rope that hauls you up and upon which you dangle thereafter. *Do not* trust a halyard, perhaps in a newly acquired secondhand boat, if it appears thin and stretched in some parts, or worn and stiff, or chafed. If you don't know its origins (eg whether it was good quality UV proofed rope), replace it at once. To tail a new halyard to an old one is simple enough (Fig 16). Starting a climb on a brand new halyard of maybe only half an inch thickness but a 2 ton breaking strength does great things for one's courage.

A bosun's chair, traditionally, is a simple plank as shown in

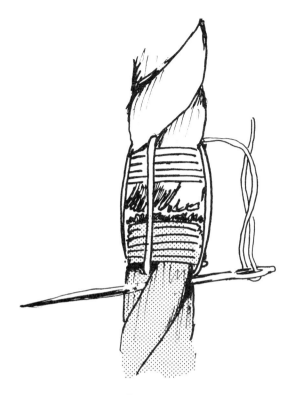

Fig 16. Tailing on a new halyard.

Fig 17; it is almost completely safe although it is just possible to slip out of it or to tip backwards if careless. The modern canvas type is certainly more secure – although perhaps less comfortable because of its tight grip around the hips and buttocks. With the chair there must be a canvas bag for tools. Don't use a plastic bucket because it will be top-heavy and liable to tip its contents upon the heads of those below on deck. Other tools may be needed while aloft; these can

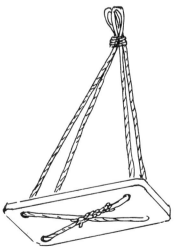

Fig 17. A bosun's chair.

be sent up clove-hitched to the burgee halyard which, being endless line, can be kept under control. Don't let anyone try hauling up a bucket of tools *on the end* of a line because there may come a stage when the weight of the hauling part of the line becomes greater than the load – and it will rocket aloft out of control.

It is wise for a climber to wear deck safety harness as an extra precaution. While the halyard may be more than strong enough, the fact that other hands will be dealing with hoisting and making fast at the foot of the mast means that the climber dangling up there is at risk should somebody fumble, and this is where a harness taken around the mast and snapped back on to itself could save the day. The length of chair slings must also be carefully considered, especially if masthead light, radio aerial or other instrumentation must be worked on. With slings rising to chin height and with the halyard shackle above that, the climber can only be raised to a height at which this point is hard up against the masthead sheave – which is far too low to be able to reach above the masthead in comfort. Send the chair aloft first (remembering a recovery line) and shorten the sling as might be needed. Remember, though, that the shorter the sling the less secure the climber will feel.

Hoisting a heavy person aloft – or any person for that matter – can be a considerable problem in any but the bigger yachts with their massive mast winches. Mast halyard winches in average sizes of yachts certainly lack the power and many lack a ratcheting facility, and often it is not possible to lead halyards back to the more powerful cockpit sheet winches. An athletic climber may be able to shin up the mast in short spurts while the bosun's chair in which he/she sits is pulled up simultaneously, but the struggle to get on to the crosstrees and above them defeats all but the fittest. There are a number of climbing aids on the boat market such as ladders of one kind or another, and also adaptations of rockclimbing and mountaineering equipment, but there is a simpler alternative.

What is needed is a powerful tackle (Fig 18), which is hoisted aloft on a halyard. Since the amount of line needed to make up a five-part tackle that extends from masthead to deck would be formidable, the tackle is made long enough in reach to take the climber to the crosstrees in one haul and then, by extending the tackle again to the masthead this time, the rest of the way up the second haul. This might involve buying some 150 feet of suitable line plus the necessary blocks. Since this tackle will also provide the yacht with the means of recovering the victim of a man-overboard incident, it is a vital piece of cruising equipment to have.

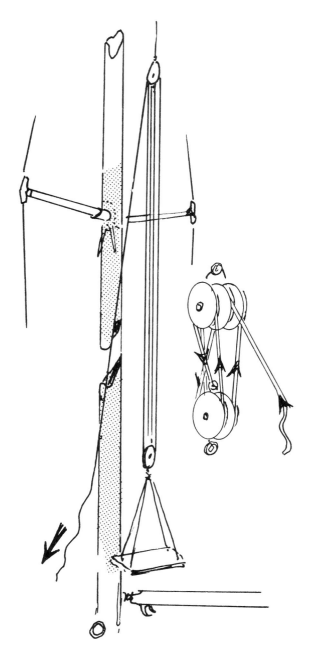

Fig 18. A powerful tackle hoisted on a halyard.

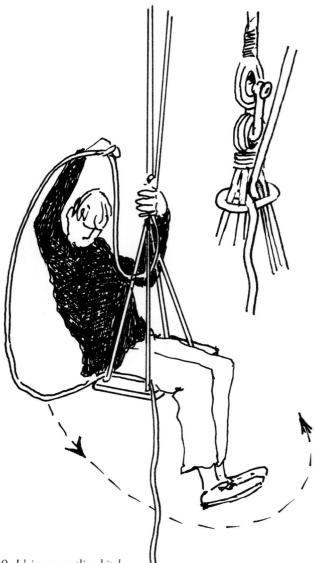

Fig 19. *Using a gantline hitch.*

In working one's way down the mast, inspecting and cleaning, either an assistant on deck is needed to lower the climbing halyard by surging it round a mast winch or the climber can lower himself. If a tackle is used, this is of course a simple matter; if the mast has outside halyards, it is also simple by making use of a *gantline hitch*

(Fig 19). In effect, this hitch is eased, thus allowing the climber to descend under control by lifting the weight of the hanging halyard tail so that the rope hitch surges smoothly.

Externally rove halyards are undoubtedly better when it comes to mast climbing in so far as the climber can assist in hoisting his/her own weight, but if the mast has internal halyards – which of course prevent use of the gantline hitch – it is quite simple to send a single block aloft to the masthead on the halyard with a gantline rove through it up and down. There is a point to note in passing – don't wear shorts when working aloft as shins and knees can take a lot of punishment and chafe from fittings and wires.

Reeving internal halyards

When a mast is standing the reeving of a new halyard is usually quite simple. A length of light signal halyard is dropped from the masthead sheave weighted by a 'mouse' consisting of a small strip of sheet lead rolled to the thickness of a pencil. Having caught this mouse at the foot of the mast the signal halyard is used to tail on the new halyard. Don't try to haul it up from the foot of the mast because the weight of the rope will impose a heavy strain on the attachment point; instead, haul the new length down through the mast so that the weight of the rope is then assisting. If the mast is out and lying in the horizontal, reeving a new halyard is much more difficult. If an electrician's 'snake' can be borrowed the job is easier, or it may be possible to use expanding wire curtain cord as a messenger, but much will depend upon the presence of wires and sound insulation inside the mast.

Whatever the situation may be, the sheaves at the foot of the mast may have to be removed – possibly even a masthead sheave as well – and there is really only one way to do this. Use a small strip of adhesive tape – such as painter's masking tape – and attach the end of a length of signal halyard to the groove of the sheave. By turning the sheave this will carry the halyard right round the circumference and, with both ends joined, you now have a means of extracting the sheave after removal of its pin (Fig 20).

Stepping the mast

If the yacht is afloat and the mast is keel-stepped don't allow the crane to lower it in if the water is at all choppy or if the boat is rolling or moving around. A keel-stepped mast that has to pass through a

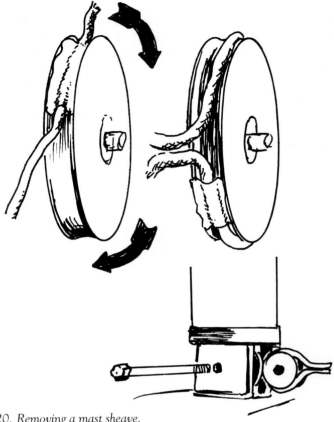

Fig 20. Removing a mast sheave.

hole in the deck can do a great deal of damage below-decks when trying to guide the heel into the step. An experienced yacht yard craneman won't even attempt to step a mast if conditions are unsuitable, but a day-hire mobile craneman unfamiliar with the job might well attempt it.

A deck-stepped mast isn't much easier if conditions are not right. It is a race to get the mast heel settled, the bolt shot home, and forestay, backstay and two shrouds set up well enough for the crane to be able to relax its hold so that the mast can move and sway around unrestricted. It all adds up to the importance of mast preparation prior to stepping.

A mast lying on deck or on trestles must be dressed ready for stepping: shrouds, stays, halyards and crosstrees in place and all leading correctly. I prefer to reeve off the halyards first, checking that

MASTS AND RIGGING

they are leading the correct side of the crosstrees and seizing them lightly to the mast with wool which will snap easily later on. If the shrouds have toggle terminals at their upper ends which fit into slots in the mast, make sure that these will remain properly engaged. They are safe enough once the wires are set up, but may need a small strip of masking tape to hold them prior to this; the risk is that they may become semi-disengaged – set up without anyone noticing and damaged in the process.

Add the standing rigging, seizing the cap shrouds into the ends of the crosstrees. Incidentally, it is a good thing to mark the wires where they locate in the crosstree slots when they are removed at laying-up time, because if the cap shrouds have been wiped down all signs of previous position may have vanished and the angle of the crosstrees when the shrouds are set up is very important. The forestay, backstay and cap shrouds are left out ready for use, the bottlescrew pins, split pins, screwdrivers and other tools are made ready for use, and you are ready for the lift. The craneman will put on a rope sling using a hitch that allows it to be recovered after the mast is raised.

Should you find afterwards that a halyard is leading down on the wrong side of the crosstrees, there is a simple way of re-leading it without the need to send someone aloft in a bosun's chair. Attach a weight – a jerry can of water, small kedge anchor or such – to the halyard and attach a heaving line beneath it. Hoist to a height of a foot or so above the crosstrees and, using the heaving line, swing the weight back and then forwards over the crosstrees, lowering away smartly and thus dropping the halyard over the crosstrees on the correct side.

Setting up the rigging

Standing rigging has more to do than simply hold the mast up. When a mast is carrying sail, stress loads assail it from many directions according to the point of sailing, and these loads vary constantly according to the strength of wind and the state of the sea. The wind strength changes constantly and the pitch and roll of the yacht imposes tremendous snatching loads on top of all. The tension applied by sheets and the violent shaking of a mast when sails are flogging set up different stresses.

Sails are cut and designed so that they assume a particular shape when properly set and they can only do this if the spars and wires along which they are stretched conform to the sailmaker's requirements; thus a mast must retain its shape and a forestay must stay tight no matter what else may be going on. A mast that is allowed to bend the wrong way in the middle and a forestay that is slack will render the best-made sails inefficient.

No matter how straight a mast may appear when at rest, it is certain to bend to some extent as soon as sails are set and sheeted home and the breeze freshens. It is *how* a mast bends that is so important, but there are wrong as well as right directions for mast-bend. For instance, if the mainsail is allowed to pull the middle of the mast aft then the head of the mast moves forward and causes the forestay to slacken. The result is a yacht with its jib luff sagging away in a great curve and the mainsail with a great hollow in its belly. If at the same time the head of the mast is sagging away to leeward, the entire sail plan is totally inefficient (Fig 21).

What we should be aiming for is a mast which, when under load in

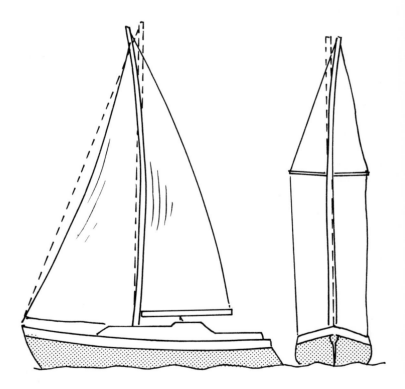

Fig 21. Mast bend.

a fresh breeze, is bending forward a little in the middle to flatten out the mainsail, while tightening the luff of the jib as its head moves aft. With the cap shrouds properly set up the head of the mast remains central, instead of bending to leeward.

Tuning the mast

First we get the mast upright. Lower the end of the boom to the deck in order to relieve the topping lift of its weight. To check that the mast is upright take the end of the main halyard down to one of the shroud chainplates on the sidedeck and adjust it so that it just touches, then carry it across the deck to the opposite chainplate on the other sidedeck – which it should also just touch if the mast is quite upright. Bear in mind that with rope halyards there will be more stretch and more catenary in the curve of the halyard, and allow for this. With the cap shrouds set up hand tight to hold the mast in this position, adjust the fore-and-aft rake.

This can either be a matter of judging by eye or, provided the yacht is floating to her marks (ie level with her waterline), by dropping a plumbob (burgee halyard plus a weight) from the masthead and raking the mast aft by no more than 6–12 inches. The backstay and forestay are set up hand tight to hold it there. We now have the mast standing upright and correctly raked and it can be seen that the four masthead wires (forestay, backstay and two cap shrouds) determine the position of the masthead; what happens to it lower down, the amount of bend and its direction, is determined by the lower shrouds and the pull exerted by the sail.

These lowers are typically either four short wires terminating just below the crosstrees, two after and two fore shrouds, or one after pair of shrouds and a babystay or short inner forestay running from below the crosstrees to the forward end of the coachroof. This babystay does the same job as the fore shrouds in pulling the middle of the mast forwards.

Begin tuning by first tightening the babystay or the forward pair of lowers until the middle of the mast is pulled forwards out of the straight by about 1 inch or the thickness of your thumb, then tighten the backstay really hard, using a spike or a large screwdriver as a lever. You cannot over-tighten by hand power alone, so don't be afraid. Tightening the backstay is also tightening the all-important forestay and there will be a lot of slack to take up. In fact, there may be some 80 feet of wire involved before you begin to really load the wire itself.

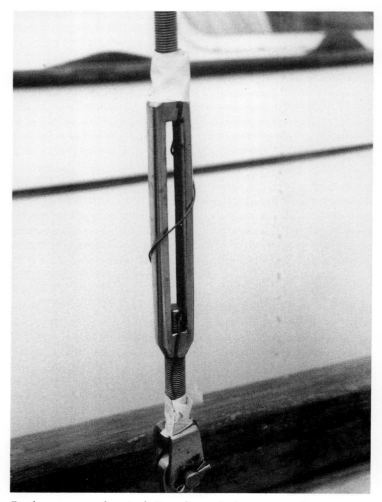

Bottlescrews must be seized. Some have a locknut system, but seizing with Monel wire to prevent accidental slackening off is safer.

Check the mast for fore-and-aft bend over its full height; there should be no more than a hand's breadth at this stage. If it appears that you have overdone things, slacken the backstay a little and tighten the forestay, remembering that when you look up a mast with your eye close to it any bend will appear to be considerable. Do not fiddle about with the lower shrouds; save this until you go sailing.

Pick a day when there is a moderate breeze, sail the boat close-hauled but not pinched up too hard, and sight up the mast to make sure that it still has that slight forward bend in the middle. If necessary, adjust the babystay or forward lowers. If there is any lateral bend in the middle of the mast, take it out by tightening the after lower shroud then tack and adjust its opposite to match. Don't overdo things with these after lower shrouds, because if they are too tight the mast may become straight in stronger winds but pulled to windward in lighter ones. Once again remember that the four longest wires hold the head of the mast in position while the lower, shorter wires pull its middle this way and that. Once all seems to be correct, lock all bottlescrews and/or seize them with Monel wire. This is of urgent importance – do not leave seizing the bottlescrews to some later date or you may forget altogether and lose your mast. When sailing, all wires are under constantly fluctuating pressures and strains, the leeward shrouds will be quite slack compared to the weather ones, and the oscillations as the boat rolls and pitches always result in threads slackening, never tightening.

Setting up a keel-stepped mast (bermudian that is) follows much the same routine, although setting that slight forward bend in the middle of the mast means that its position in the deck opening is critical. In general terms the mast should be closer to the after end or side of the hole than to the forward end while standing evenly spaced laterally, and thus when the mast is properly set up this spacing will accommodate the slight forward bend. This deck opening – the mast partners – is fitted with tapered wedges, wooden ones in the case of wooden masts but non-tapered rubber ones for an alloy mast. Inserting these rubber wedges can be a struggle. Lubricating with liquid soap will help, but pulling the mast bodily aft by a fraction may also be necessary and this can best be done by taking a rope from the mast to the two sheet winches back aft and heaving away on one of them.

Gaff rig

The keel-stepped mast of a gaff rigged yacht may be either a short mast fitted with a separate pole topmast, which can be hoisted into place and secured at its heel with a metal fid, or it can be a single spar. In either case, setting up is a more simple matter calling for getting it upright laterally and with the rake desired. The jib will almost certainly be set flying and tensioned by means of tackles rather than depending on the tension of a forestay and the staysail will be setting on a short forestay which is no problem to tension; running backstays hold the mast straight against the pull of the jib. Wooden solid masts are never over-stressed, and though they should never be allowed to slop around in the deck partners it is an unnecessary strain for an elderly boat to have rigging set up twanging-hard; chainplates, stem-hook and garboard strakes (planks) can be damaged as a result.

8

Plumbing, winches, electrics, guardrails

The belief was that the fewer holes you bored in the hull of a boat the safer it was, but nowadays a fairly typical 32-footer can have up to ten of them, including through-hull instruments, drains, inlet pipes and so on. Apart from the holes that take instrument heads or transducers, all other holes will be fitted with seacocks from which plastic hose carries or drains away water. All that prevents any of these holes from sinking the boat is turning off the seacock, and when the seacock is in the on position safety rests entirely with the jubilee clips which secure the plastic hose to the spigot of the seacock. When we consider that these pipes are often lying exposed in the bottoms of lockers and that objects are constantly being tugged out of these lockers, it is little short of a miracle that loosely clipped hoses are not being dragged off as a regular mishap.

Another oddity is that the majority of seacocks are almost totally inaccessible because of the boxing in and general accommodation carpentry, which means of course that seacocks tend to get little or no maintenance. I think that it should be a basic priority of boatbuilding that all seacocks, as well as all sides of the engine and other such vitals as chainplate bolts, should be either left accessible or only boxed in with removable panels.

Many wise owners make a point of turning off all seacocks when the boat is left for a while and turning off toilet and sink waste seacocks routinely after each use; this not only guards against accidents but ensures that the seacocks remain easy to turn and don't seize up from lack of movement. Whatever the policy, the thing to bear in mind is that whatever its function a hole is a hole – and thus potentially capable of sinking you.

Undoubtedly the hose and its jubilee clips is a glaring weakness. Careful owners double-clip every hose attachment at the seacock, but the other end of the hose is just as important and boats have been sunk by a disconnected inner end acting as a siphon. Be alert to hoses also with what you stow in a locker occupied by a hose; a small kedge anchor needed in a hurry for instance is an accident in the making.

In fitting out, examine every seacock and test jubilee clips with a screwdriver. If a seacock lever won't budge and it is in a permanently 'on' position, it must be made to budge even if this means cutting away cabin carpentry. It must be stripped down, greased and reassembled and the fitting-out programme should stop until this is done. Sometimes an overnight soaking in penetrating oil followed by a light tap will start a lever moving. Don't leave matters at that though – continue by stripping, cleaning and greasing with pump-grease. In frosty weather be very cautious of administering light taps to any metal parts; frost tends to make metal brittle and easily snapped. In the case of taper plug valves which may turn and then stick, the use of valve grinding paste and gentle working to and fro will usually remove any hard spots.

Wheel valves are another matter. If still immovable after thorough soaking with penetrating oil, *do not* try to force them or the spindle may snap. Dismantle the whole valve if possible. Use only spanners that are an exact fit though because the flats of a nut, being made from soft metal, are easily stripped and there is then no way of holding it – except by using a large Mole wrench which will ruin the nut completely. The last resort of course is heat. Be very wary though. Surround the area with soaked rags and use a small aerosol blowlamp with a finely adjusted flame. The aim is to heat and therefore expand the nut, thus breaking the verdigris that is preventing it from being unfastened.

Corrosion is a hidden danger affecting all seacocks. If for some reason electrolysis is present, as a result of an unwise mixture of metals in seawater or the escape of electric currents from the yacht's circuit, it will most probably be the bolts holding a seacock fitting to the hull that sustain damage. Look for signs of white powder corrosion and for a reddish hue affecting brass alloy. Bolts afflicted become brittle. Apply force by tightening down on the nuts. Certainly replace bolts if there is any chance that they are brittle, but, more importantly, find out how it has happened and remedy it.

Removing such bolts is often more difficult than it sounds. A heavy weight held close to the bolt head on the outside plus a punch in the inside and a hearty thump with a heavy hammer usually works but – and it is a big one – it takes an expert to hit hard and true in the narrow confines of a locker. Seacocks are important and potentially lethal, so don't bodge the job. If you can't cope, get an expert.

Fresh-water plumbing

Internal plumbing is a different matter. The fresh-water supply may be contained in glassfibre tanks moulded in when the boat was built. Such tanks contribute to the strength of the hull and they neither leak nor corrode. In new boats, however, there is all too often a lingering taint of styrene which takes months to eradicate no matter how often the water is changed. It is a result of the resins not having cured properly and it is better to drink bottled water and keep on flushing meanwhile. In older boats this styrene flavouring may be a sign that osmosis is occurring within the tank. In such cases, only cutting out the top of the tank and installing a flexible plastic one inside is really effective. This styrene taste may occur briefly every year in some cases, disappearing after a week or two. It must be dealt with because the chemical is mildly poisonous and a common cause of poor health on board, certainly one cause of sea-sickness.

Although expensive, stainless steel water tanks are vastly superior to all others, but even these must have removable access hatches to allow for interior cleaning. Wooden tanks are not unknown and quite serviceable provided there is cleaning access. Galvanised steel is common in older boats, often tarred inside, and the same proviso holds good. The flexible plastic water tank is also a common feature; it is excellent in every way *provided it is properly supported*. Remember that 30 gallons of water weighs around 300 pounds and that this weight, constantly surging and sloshing around, rubbing the plastic tank against the hull, rapidly chafes it away. A flexible tank therefore must be supported, padded and lidded to restrain it.

Bacteria

Tanks and piping are prone to infestation by bacteria of various kinds, which can range from harmless to smelly, mildly unsettling to the stomach, or downright poisonous. There are various filter systems that can be fitted and that claim to remove all impurities, but this is not as simple as it sounds. Not all such filters are as effective as their makers claim, some even harbour harmful bacteria, and all rely upon having a cartridge regularly changed.

A common feature in many boats and upon coming aboard after the boat has been left for a period is the foul smell of rotten eggs which is released as soon as the toilet is used. There is in fact a particular bacteria that causes this; it breeds in pipes that are transparent and allow daylight to enter. Though it is relatively harmless, the smell is quite disgusting. Changing to non-translucent pipes or taping over existing pipes sometimes prevents the odour.

During fitting out the water tanks should be washed through and disinfected. All piping including the deck filler pipe should be flushed and disinfected at the same time. The method is to flush out and then refill the tank, adding a water purifier such as Puriclean at a rate of 80 grams per 50 litres. It is best to mix this in a strong solution and add it via the deck filler pipe so that this also is treated. Leave the system full for at least four or five hours, or preferably overnight, and then pump out using all outlets. Refill and empty again, refill once more, and then try tasting the water.

The worst source of infection is often the marina hosepipe. Left coiled up and in the hot sun it is a breeding ground for bacteria, and before using it the hose should be allowed to run free until the water feels cool. Better still, if it is possible to unscrew the hose at the tap, take a funnel and pour a good strong dose of purifier through the hose before filling your tank.

Whether you have an all-electric fresh-water system aboard or prefer to waggle a pump and avoid taking a massive dose of current out of the ship's battery, the hose in use needs to be checked. In general terms, the more flexible a hose is the less suitable it is for carrying potable water – it will probably have a strong plastic flavour. Even some of the recommended grades of hose can taint the water, but the best type is the black polythene hose; this denies daylight to the bacteria that may be present and thus restricts growth. Polybutyl is another suitable hose material. While it may be asking too much of an owner to replace all existing hose with an approved type, this should be borne in mind if purification and the use of a filter still leaves the water tainted and unpleasant either when drunk alone or in tea or coffee. Consider adding a smaller tank for drinking and galley use only, if you do not want to change all the piping.

Pumps and toilets

All manufacturers of pumps and toilets supply spares kits which include spare gaskets, washers and valves. Make a point of keeping a kit aboard for every pump so that running repairs can be carried out while cruising away from home. Valves and washers become worn after a few seasons and gaskets are often ruined when stripping down a pump.

The bilge pump will probably be a diaphragm type and, provided the inlet hose has a 'strum box' to prevent small objects from blocking the pump valves, little can go wrong. Strip the pump down though and check the state of the neoprene diaphragm. Although this pump may have nothing to do in a plastic hull and, barring a drip from the sternshaft gland, little bilge water to need removal, it remains the most important pump on board. A boat can be holed on a rock or a skin fitting can fail, and it is then that the bilge pump becomes life or death to the boat. Check that the end of the intake pipe is as low in the bilges as it is possible to get it. Plastic hose has a tendency to curl and this may raise the pipe end above the bilge water level. A strip of sheet lead clamped around the pipe end will hold it down. If you cannot get hold of a proper strum box, a wad of galvanised chicken wire rolled up and forced into the end of the hose does very well.

Marine toilets will doubtless become obsolete in time, under pressure from environmental considerations, and all yachts will have either efficient chemical toilets or large holding tanks. Unless there are to be shore facilities or a pump-out barge service in every yacht harbour and marina, holding tanks will merely mean holding until they can be discharged at sea. Meanwhile, we have the mechanical toilet to be serviced.

Types vary from a vacuum-operated model with a pickle-jar lid to the eject-and-flush sort of which there are many versions, some with single lever operation and others with separate flush and eject pumps. In many cases, toilets flood over if the intake valve that admits seawater for flushing is left open; some are fine in harbour but flood rapidly on the toilet-side tack when heeling under sail. A flooding toilet is a liability because it relies upon human memory to turn off the cocks after use: there's bound to be a day when someone will forget. Check then to see if the delivery pipe can be lengthened, extended in a curve upwards well above heeled waterline and back down to the toilet. Remember though that such a pipe can siphon.

Since most yacht toilets are tucked away in a tight corner, working on them is difficult and gymnastic; a first priority then is to ensure that the whole thing can be unbolted and taken on deck to be worked on. This usually means leaving room for a spanner to reach the holding down bolt nuts and heads, and also the pipe unions for both flush and soil pipe. Assemble with well-greased nuts. Plastic pipe once clamped tight by jubilee clips adapts to the shape of the spigot and becomes very difficult to move. Pour boiling water on the pipe to soften it and, wearing gloves, wrench it off before it can harden again. Remember that the toilet is the hardest-worked piece of machinery aboard; engines and winches also work hard but they are seldom treated as harshly as the loo.

The usual trouble is a blockage of a valve flap – perhaps by some unwise person dropping some insoluble object down the toilet. I read a short verse behind the door in one boat which read: 'What you have eaten, so can I. You don't eat peach stones, I won't even try.' Having replaced a faulty valve, reassemble the pump with a smear of Vaseline on all gasket surfaces and tighten groups of bolts evenly. In the case of a leaky pump rod where water squirts up past the washer on the delivery stroke, a new washer may not always cure the problem – which is caused in the first place by users jiggling the handle rather than giving it a full length stroke, and hence the middle of the pump rod wears thinner than the top or bottom. A new rod is the only answer.

Galley pumps are much the same; they are often difficult to remove and consequently are rarely maintained. Many have a large screw collar underneath the galley work-top, well boxed in by enthusiastic carpenters and hence totally inaccessible. Since this collar is plastic it cannot be treated roughly. Penetrating oil or diesel oil, if they can be applied, may loosen it; failing this, you may have to snap it by levering with a chisel and sending away for a new collar.

Winches

Do not begin dismantling a winch until you have seen a diagram or cut-away drawing of the winch and the mechanics involved. All winch manufacturers supply maintenance instructions free.

The work is usually simple Meccano mechanics: a matter of stripping down in stages and noting what went where and which side up. Parts are washed in paraffin with a small stiff brush, then dried and reassembled with the recommended grade of grease. Use only the recommended grade; failure to do so may result in a winch that by mid-season is choked with emulsified muck which jams up the pawls and stiffens the whole action.

When stripping down any winch *in situ* it is wise to rig up some sort of surround for it – such as a large cardboard box adapted to provide walls on three sides. The reason for this is that winches tend to have tiny high tensile springs which go winging out into the blue before you can catch them. There may be a circlip to secure the top plate on the barrel, and each pawl will have its own tiny coil spring. Spare pawl springs should be carried aboard.

Some sheet winches are much more complex than others. The basic single ratchet winch offers no problem, but the multi-gear double ratchet types must be dismantled with system and planning. Anchor windlasses are notorious for working stiff after a winter's neglect. If you have a new windlass installed, make a point of maintaining it to maker's recommendations annually and keep it under a spray cover.

The electrics

The salt-air moist environment of a boat is quite the worst possible for electrics of any sort. Not all equipment and fittings are designed with this in mind – some items aboard boats are really caravan fittings marinised by different packaging. In fitting-out terms much depends on how trouble-free the electrics have been in the past, whether particular fittings have given regular trouble, and so on.

Most problems stem from loose or corroded contacts; indeed, any electrical joint, connection or terminal that is not actually sealed against damp must be suspect. Beginning with the main battery, test every connection. Look for any sign of white corrosion or verdigris and renew and Vaseline any connection affected. Condensation can play hell with electrics too and cause shorting in extreme cases. Bulb contacts often fail as a result of corrosion, and let us not forget that galley lights that are exposed to steam can become a problem. Look to the fuse box – if it isn't really accessible make it so. Failure of navigation lights in busy shipping could be a very serious matter, more so if the fault lies in a blown fuse that cannot easily be replaced.

The main battery or batteries should be taken ashore to a garage and given a proper check to establish that they are capable of taking and holding their charge. Perhaps the majority of small yachts run on batteries that are permanently on half-charge and the proliferation of electrical instruments added year by year makes ever-increasing demands on boat batteries. Check capacity against cruising current consumption. Quite frequently a new yacht bought complete with one battery becomes increasingly overloaded with equipment while the battery steadily deteriorates. If you suspect that this may have happened, it is time to consider not only the consumption/capacity aspect, but also the size and suitability of wiring and fusing.

Navigation and other outside lights must be absolutely watertight. Examine closely for signs of cracking in the plastic cases. If they can be dismantled without damaging the gasket or water seal, do so to check that bulb contacts are good but do not touch the halogen-type bulbs with bare fingers. Reassemble with a smear of Vaseline or petroleum jelly on all threads and gaskets. Deck sockets can be a common cause of fuse blowing – usually because the protective cap is left off or replaced cross-threaded, this allowing water to enter. Remember that poor contacts can mean current drop and electrical leakages can mean electrolytic corrosion of skin fittings, propeller and other such vitals.

Deck leaks

In theory a glassfibre deck doesn't leak; in practice there are several potential weaknesses – such as the joining of hull to deck, the sealing of windows into the coachroof moulding, chainplate fittings which pierce the deck, and of course hatches and ventilators.

Should a deck-hull joint fail while a boat is still new it is a matter to take up with the builder without delay. In older boats the appearance of one leak here may herald a crop of them and extensive action may be called for as access to the deck joint will probably be hidden by the inner moulding.

In fitting-out terms, though, leaks are more likely to be easily dealt with. Check wherever you can see the state of deck-piercing fittings, and if a chainplate nut for instance is hidden behind an interior moulding it is well worth cutting an access to it. If a shroud plate, stanchion bolt or sheet track is leaking it must be slacked back and hardened down on a good luting of a suitable mastic. It is no use trying to smear mastic around such bolts.

Windows give more trouble from condensation than from actual leakage, although when this appears it can only get worse if ignored. Windows may be fitted into a neoprene moulding within the window opening, and if the area of plexiglass is large the flexion of the whole pane can in time cause failure of the moulding and complete renewal is the only answer. On the other hand, a window may consist quite simply of a pane of plastic that is larger all round than the window hole and secured on a mastic bed by a large number of screws which may or may not hold in place a decorative outer trimming strip. In either case, never bodge. Remove the whole pane, rebed, and replace with larger screws if the originals are no longer holding.

Hatch leaks are usually the result of poor design, although the foam plastic gasket often fitted may have perished. Check that the forehatch can be tightened down from below so that it is compressing the foam gasket evenly all round. Deck ventilators should be able to withstand a bucket of water being hurled at them; leakage, if any, may be due to blocked-up drain holes. Cockpit locker lids are another potential source of water finding its way below for they often allow direct access to the bilges or even to the accommodation at their forward end. It should be possible to seal such lockers securely by means of hasps on the outside. Remember that a yacht under severe stress of wind and sea, rolled over on her beam ends or further, could be put into danger by the huge cockpit lockers of today becoming swamped.

Guardrails

The philosophy regarding guardrails is a curious one. Although the reason for having them is to provide a safety fence to prevent people from falling overboard and thus they should be strong enough to withstand the weight of a heavy man, in normal practice we do all we can to avoid touching them. We pretend they are not there at all – and the person who leans against the rails, drink in hand, or helps himself aboard by them, labels himself a novice of the first degree. That guardrails and bow and stern pulpits were unknown a short generation ago is not always appreciated, but nowadays, as a vital part of every yacht, sailing at sea would be unthinkable without them.

Inspecting guardrails at fitting-out time should mean taking a large screwdriver and testing every stanchion fastening as it is quite common for the hidden nuts to slacken, perhaps due to engine vibration, and the resultant movement then steadily gets worse. The wires, if plastic covered, should also be stainless and, if not, they are highly suspect. Lightly galvanised and plastic-coated wire which chafes in the stanchion heads can begin rusting internally and is thereafter liable to part without warning. Check also that the stanchions are not bent as they are very vulnerable to knocks when going alongside.

The end terminals of guardrail wires – be they spliced or Talurit eyes, mechanical terminals or whatever – must be examined carefully. The usual arrangement is to shackle the forward ends and make up the after ends to the stern pulpit with a lacing line; this not only breaks the electrical circuit around the ship caused by metal-to-metal connection (and a cause of radio distortion), but it ensures that the wires can be freed with the single slash of a sharp knife. In the event of a man-overboard crisis this could be critical in terms of

FITTING OUT

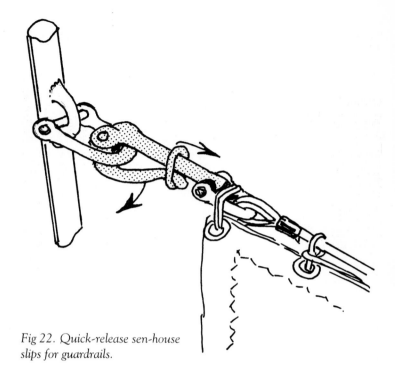

Fig 22. Quick-release sen-house slips for guardrails.

recovering him back aboard. Unless you bought and rigged this lacing yourself, do not trust it. A previous owner may have used a cheap bit of line which is not sun-rot proofed and it may in fact be lethally weak.

It is possible to buy quick-release sen-house slips for the attachment of guardrail wires, but do be sure that everyone aboard knows how to use them (Fig 22) and that they are not too stiff for weaker fingers to operate. If the wires are set up really tightly, it is often the case that these slips are very difficult to release.

It is sometimes only necessary for the lower of the two wires to be capable of being released, if this leaves adequate room for a rescued casualty to be dragged or rolled aboard underneath the upper one. If there is a canvas or plastic name dodger rigged this also must be attached for quick release. There is a simple and cheap method of doing this. The use of a large split pin acting as a peg and with a lanyard attached means that guardrail wire or dodger lashing can be released instantly. In setting up this arrangement the actual tensioning of the guardrail wire is done by means of a lacing at its forward end after having inserted the peg at the after end.

When lacing a dodger to a guardrail use this locking hitch; it will remain taut even if one end becomes loose.

Incidentally, the name dodger has been found to become a hazard in cases of severe weather and knock-down when a yacht is laid on her beam ends in heavy seas. In such an extreme, the dodger having been forced under water then acts as a brake when the yacht attempts to roll upright again. For this reason and because a dodger may need to be let go for the purpose of recovering a person back on board, it is better to secure the lower edge to the rail with strong rubber cord hooked over small brass or stainless thumb cleats. Failing this, a light and easily snapped lacing can be used.

9

The auxiliary engine

This subject will be dealt with fairly briefly since the matter of engines is not suited to wide generalisations. There are so many sizes and types of auxiliary that only a full-length book could hope to cover them in any detail. For the purpose of fitting out then, we have to assume that the engine is healthy and that it was running satisfactorily when the boat was laid up; if this is not the case, then the attentions of a marine engineer are needed.

If the engine was properly laid up in good running condition there should not be a lot that needs doing at fitting-out time. It is probably a safe bet though that very few engines are maintained in full accordance with the maker's handbook; this is due in part to the fact that human beings are rarely perfect and also to the fact that most engine installations in yachts are designed to make proper maintenance difficult. If only builders would surround engines with easily removable panels!

There is one school of thought that holds that when an engine is running smoothly it should be left well alone – apart from feeding it with fuel, oil and cooling water. While uninformed fiddling with engines is not to be advocated, it is equally as bad to ignore such matters as the changing of oil, fuel and air filters and the regular replacement of the water pump impeller.

Diesels

A healthy engine properly winterised should give no bother when starting up in the spring. Winterising should consist of the following: With a fresh-water supply replacing the salt intake, the engine is run until warm; after this, the sump oil is drawn off and replaced by the correct grade of inhibiting oil. The engine is given another short run to distribute it, and the water supply is substituted by a 50:50

THE AUXILARY ENGINE

mixture of fresh water and antifreeze. Fuel, oil and air filters should be replaced and the water pump impeller either removed for storage or smeared with petroleum jelly to stop it from sticking while lying idle. The batteries will have been on full charge and in store all the winter. The engine should give a preliminary splutter when started up and then settle down to smooth running.

However, none of us is perfect. We may have cherished the engine, but the fuel supply may have received scant attention. Ideally, a fuel tank should be left empty and well cleaned out or topped up brim full; in both cases the aim is to limit the danger of winter condensation inside the tank. An empty tank can be drained prior to refilling while a full one leaves no room for condensation. The water drain on the primary filter should be opened prior to starting the season in any case, but if the tank was left partly filled there may be a layer of water beneath the fuel to be got rid of. It is better to drain or siphon out the tank completely, stirring up the last dregs so that any sludge is in suspension and drawn off, then fine-filter the fuel back again.

This attention to fuel purity is vital for diesel engines. Basically they are simple machines, but clean fuel, clean tank, clean primary filter, clean or renewed filter units are of primary importance. Attending to these matters will of course mean that air will invade the fuel lines and the system will have to be bled from tank to injector. The user's manual for each engine details this procedure and owners should become familiar with it because a tank becoming a bit low, plus a bit of a tumble in a seaway, can result in an airlock and the need for some rapid bleed work.

Check that the gear oil is changed or topped up and check also the gear shift system. This will probably be the single lever combined throttle and gear arrangement which is cable operated. In some cases, this cable must be allowed freedom to move bodily as the lever is operated; if this cable movement has been restricted in any way the wire inside the cable cover may have become fatigued to the point of breakage.

Petrol auxiliaries

Since it is rare to find a new boat fitted with a petrol motor, the norm is an older boat with an ageing petrol auxiliary. It helps to know the history, particularly if the boat has had a succession of owners.

The cause of starting problems with petrol engines is almost always traceable to damp electrics, poor contacts, dirty fuel or a stopped jet. If the engine was running smoothly prior to laying up there is no point in messing about with timing or similar adjustments, but it is always sensible to start the season with a new set of plugs correctly gapped.

Four-stroke engines usually show some sign of life, perhaps breaking into an uneven gallop on three pots. Shorting out each plug lead in turn with an insulated screwdriver soon indicates which one is failing.

If the engine is reluctant to start, be very, very careful, have fire extinguishers handy, and open up the engine box for maximum draught. It may be that the fault is electrical, in which case the more you crank the engine over the more raw fuel you may be pumping. If any raw fuel spills, perhaps from a faulty float chamber, stop work and wait for the fumes to dissipate.

Two-strokes

In the case of two-strokes start off with new, correctly gapped plugs, but first remove the old ones and pour in a little oil as with the diesel and turn her over by hand. Excess oil may need to be drained from the crankcase.

Examine the contact breaker points and fit new ones if pitted and clean fuel filters – better still renew them to start the season. Again, fit a fully charged starter battery and check over the cooling pump. If stale fuel remains in the tank this should be drained and fresh fuel of correct petrol–oil mixture should be provided. Tank condensation often results in quite considerable amounts of water accumulating in the tank and filters. With the oil levels checked and correct, go through starting procedures and run the engine up to normal temperature. Failure to start is usually due to an electrical fault; check the plug, which may be either damp or bridged, replace it, and try again. If the engine was running satisfactorily before lay-up do not embark upon an experimental round of adjustment to settings – two-strokes are notoriously temperamental. If compression, spark and fuel mixture are correct it will probably go away with a roar just at the time when you are giving up hope – thereafter it will probably behave itself.

10

Dismantling stubborn fittings

Consider first what harm you may do if you attempt to remove a seized bolt, nut or other threaded item and it snaps off short. Don't attempt to use excessive force on anything you are not sure about (eg it is easy to assume that a particular fitting is assembled in a particular order whereas in reality it may differ from your expectations). Corrosion of one sort or another is mainly to blame for a seized-up screw – but remember that corrosion may also mean that it has become brittle; plastic nuts and threaded parts can also become brittle with age or exposure to sunlight.

Wood screws

Brass screws in a salt environment tend to dezincify and become brittle. Look for the reddish hue of bare metal and if the screw refuses to budge proceed with great care. Penetrating oil applied in advance will soften the corrosion in the screw threads and a single firm blow to the handle of the screwdriver will jar the threads and thereby loosen them. It is important to pick the biggest screwdriver that the head of the screw will accept and, if the screwdriver blade has become rounded with use, file it square and flat so that it fits the screw slot exactly.

The trick now is to induce a minute degree of twist to the screw while ensuring that the screwdriver blade *remains in the slot*. If the blade slips out of the slot the screw head will probably be damaged beyond hope. Get the full weight of your upper part of your body poised directly above the screwdriver, and lunge down on it while simultaneously imparting a minute twist. This is a particular knack which, once mastered, is almost always successful – if not at the first lunge then at the second one, and after another tap with a mallet to jar the threads. You will feel the first faint movement; repeat the lunges and be content to proceed slowly in very small stages.

In some cases, particularly with large screws, it may be worth trying with two people. While one person applies their full weight, the other, with a Mole wrench clamped to the screwdriver flats, imparts a very small twist. While this method often works, the chances of causing damage to the screw are far higher.

With a row of screws to be removed, tackle them randomly and be content to feel movement on each before moving on to the next. If you begin to move them in succession, the strain becomes concentrated on the remaining screws which may then prove even harder to shift. The oft heard remark that it is always the last screw that causes the bother has a good sound reason behind it.

Impact screwdrivers or screw removers, depending upon how you set them up, simply reproduce mechanically the lunge/twist movement described above. With the impact screwdriver though, it is the hearty smack with a hammer that both jars and twists the screw. Usually the versions on sale only have screwdriver bits that are suitable for small screws. Remember also that if the bit is not held firmly in the screw slot it will do a lot of damage.

Nuts, bolts and studs

Penetrating oil must be allowed time to work, therefore apply it a couple of days in advance. Or a diesel-soaked rag can be left over a stubborn bolt for a week prior to tackling its removal. Find a spanner that is the exact fit for the nut – a ring or socket spanner if space permits. If there is even the slightest rock on the spanner, once maximum force is applied, it will almost certainly ruin the flats of the nut. This is naturally even more important if the nut or bolt is of a soft alloy while the spanner is made of high tensile steel.

Be very cautious about the use of force on stubborn threads during frosty weather when metal is more brittle. A single hammer blow on the end of the threaded part may jar the threads initially, but don't over-do things. Try to apply a steady and increasing force even to the extent of getting a lever to work on the spanner, or sleeving a length of pipe over the socket spanner handle to increase the power. If all this fails you may be certain that any greater force or more drastic hammering will probably sheer the bolt.

Heat is the next resort. As mentioned earlier, surround the spot with wet rags, making certain that a naked flame cannot ignite any flammable vapours that might be present, and always have an extinguisher handy. Use a small pointed flame and apply it evenly all round the nut or female part of the threaded pair – the aim being to cause it to expand and break the thread hold. If penetrating fluid has been used, the fumes released by applying heat will be toxic – so be careful. Sometimes very little heat will be sufficient, but in other cases it may require repeated heatings of both bolt and nut, or screwed stud and surrounding metal. Success may come after the work has cooled off and you have almost given up hope.

Nut and bolt turning

Sometimes a nut cannot be unscrewed because the bolt is free to turn with it and there is no access to the head of the bolt. Provided that the nut is not immovably corroded to the bolt it is often possible to hacksaw a slot across the top of the nut and the head or end of the bolt. This then allows a small screwdriver to be used on the end of the bolt while a spanner is applied to the nut.

In the event of a stubborn nut being in such a position that even an open spanner cannot be used, a sharp cold chisel can be used to cut a deep nick in the side of one of the flats. Take a good square-ended punch and a heavy hammer and smack it hard; the lateral blow should jar the thread and start the nut turning. Never merely tap a punch or something that needs jarring – weight and the inertia is what matters.

Stainless steel threads that are reluctant to move must be treated with special care as the high friction that is generated by forcing them to move can produce a 'welding' effect which can seize the threads. Any further attempt to move them by force will merely result in the part snapping off. Unless stainless parts are unscrewed gently, by easing them to and fro, lubricating them and allowing them time to cool off between times, there is a strong risk of failure.

11

Bosunry

This can call for a good deal of skill and knowledge in the case of traditionally rigged craft, but much less with modern ones. A basic kit of rigger's tools is needed and, in the list below, an asterisk indicates those required by the owner of a small modern cruiser. Much depends upon the owner's ambitions. A man with a gaff cutter to maintain will know how to splice wire and rope as well as many other smaller skills not needed by the modern boatowner.

Wire cutters*
Cold chisel
Ball peen hammer
Lead hammer
Pliers (long nosed)
Spike
Rigger's vice or hand cramp
Adhesive tape*
Splicing (hollow) fid*
Mole grips or parrot beak pliers*
Serving mallet or board

Sharp knife*
Wooden fid
Sewing palm and selection of needles*
Sail hook (bench hook)
Brass eyelet punch and dolly kit. Eyelets of most suitable size
Bee's wax
Selection of waxed threads (Terylene)*
Ball marline, hank of codline

This is not comprehensive and there may be need for two sizes of wooden fids, eyelet punches and rigger's cramps according to the size of the boat and complexity of rigging. A word or two about items on the list is in order.

Wire (end type) cutters and cold chisel are for wire splicing work; also, the lead hammer is used for tapping the splice into shape afterwards without marring its galvanised surfaces too much. The rigger's vice or cramp holds the wire firmly and evenly around the metal thimble while splicing, and the hollow fid lifts the strands leaving a gap through which to push the ends. The wooden fid (like a long, tapered and pointed bottle-neck) is used for fitting cringles or rope eyes to a sail and the sail hook is a small steel hook on a swivel and attached to a length of line. When sewing canvas this is hooked in behind the line of stitches and secured to the bench or stool so that the work can be pulled against it, keeping the seam taut (Fig 23). The eyelet punch incidentally is very useful in any boat as it allows awnings and canvas dodgers to be made or repaired.

Fig 23. Keeping a seam taut.

Bosunry in modern craft

This will be mainly limited to splicing in synthetic rope and applying seizings and whippings – and occasionally servings. These must be defined. A seizing is a binding with small stuff which holds two parts of a rope together, thus a seizing might be used to hold a metal thimble or liner into a spliced eye which is a loose fit or a wind telltale might be seized to a shroud. A whipping is simply a binding to prevent a rope's end from unlaying and a serving is the sort of binding you might see on a cricket bat handle; it is applied under as much tension as the thread will safely stand.

In the modern craft there will also be a certain amount of improvisation in the use of shackles and blocks. A boat newly bought will be rigged in working order, but most owners soon see things that they feel they could improve on or alter and the eternal juggling with shackles begins. In the main, the modern stainless steel bent strip type of shackle is less adaptable than the older galvanised type. What must always be borne in mind is that every shackle is designed to be used in a certain way and to bear a certain maximum load. The temptation to 'find a shackle that fits' irrespective of its size and type must be resisted.

Splicing and worked eyes in synthetic rope

The sheets and halyards will be either three-strand laid rope or a form of plaited rope. Synthetics are harder to splice and to work in general than natural fibre ropes as a result of their fuzzing out nature and the fact that when a strand of natural fibre rope is unlaid it retains most of its twisted form; synthetic rope, in contrast, tends to become limp and to unlay itself.

The long and short splices for joining ropes will not be described here because they are so seldom used. A warp might be short spliced to cut out a damaged portion or an extra length might be short spliced to the tail of a halyard which is too short by a couple of feet, but that's about all. Even here a long splice should be used, but long splicing synthetics, because of the reasons stated, is tricky and liable to result in a very weak splice unless done by an expert.

The eye splice on the other hand is constantly needed. The three-strand splice is simple, but the eye splice in plaited rope is apt to be more difficult to carry out neatly – although there are simple methods that we will examine which are equally strong.

Splicing in three-strand

The sketch (Fig 24) shows the principle of the thing. The size of eye is decided, the main standing part of the rope is opened up, and the three strands of the rope's end are inserted and worked along in an under-and-over fashion. It helps greatly if temporary seizings are made just below the point at which the rope is opened up and on the rope's end at a point beyond which you do not want the strands to unlay; this also keeps the finished eye to the size desired. Take the middle strand of the rope's end and pass it from right to left under the middle strand of the opened-up rope. Next deal with the left-hand end similarly and finally take your right-hand end and tuck it up under the right-hand strand so that one end is coming out of each hole. Now take any of the ends and go over one and up under the next strand; repeat with each end. You should put about four tucks in each before tugging all tight and, placing the splice on the deck, roll it under-foot to snug all down before trimming off the ends.

Splicing in plaited rope can be either the sew-and-serve method or the 'flemish eye'. Alternatively, there is a more complicated but far neater splice for use with double plait – this is a rope consisting of an inner plaited core and an outer plaited cover as opposed to the plaited cover over straight laid inner strands or threads.

Flemish eye (Fig 25)

Unlay the rope plait for about 30 diameters and, with a razor blade, shave the resulting bunch of threads to a tapered point. Next take a short length of cardboard tube, or make one roughly to the diameter

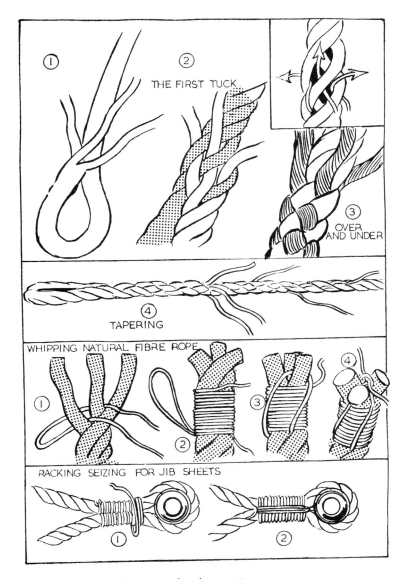

Fig 24. *Splicing, whipping and racking seizing.*

of the desired eye and make a series of slits in each end, in opposite pairs. Stretch short lengths of waxed sail twine from slit to slit. The tapered and unlaid threads are now divided equally and a stopping (or two) put around the whole bunch below where the eye is to be

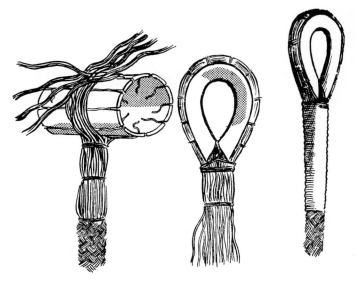

Fig 25. Making a Flemish eye.

formed. The divided threads must now be taken around the tube from left to right alternately (see sketch) and seized together below the tube at the 'neck' of the splice; tug each thread tight. You can now pull the ends of thread loose from their slits and knot them together to hold the threads which are forming the eye and, when complete, remove the tube. At this stage a serving of thicker thread is bound round the eye part, the thimble is placed in position, and a tight seizing is applied to the neck of the splice to hold it. The rest of the splice – the tapered threads lying along the standing part of the rope – must now be served over tightly (serving is dealt with separately later).

Sew-and-serve

If time permits it is better to taper the rope's end and bee's wax it before beginning to form the eye and, in the case of making up a pair of jib sheets, this is really essential. If the rope is not tapered the butt end of the rope forming the eye will snag up repeatedly against the lee shrouds when sheeting home. Having made the eye leave it soft (ie don't fit a thimble in it), and with a sail needle and waxed thread sew along the two parts and back again (Fig 26), then along each side. End with a tight seizing at the neck of the splice and another one over the tapered area.

Fig 26. Sew-and-serve.

Seizing, serving and whipping

When applying a seizing to two parts of a rope, first overbind the doubled end of the twine (Fig 27). Having put on the necessary number of turns, the end of the twine is passed through the loop which is then hauled up inside the seizing by heaving on the original end. Heave taut on both ends of twine and trim off.

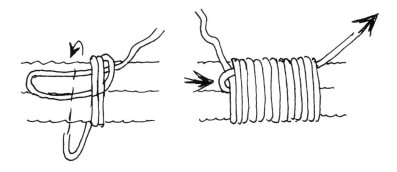

Fig 27. Applying seizing to two parts of a rope.

To seize an eye in a length of rubber cord, first trap the twine under your thumb while overbinding as above; this allows you to haul it taut compressing the rubber prior to adding the rest of the turns. It will be difficult to align each turn next to its neighbour because of the bulge of the compressed rubber, so use fairly thick twine. The end of the twine, which will be pulled to draw the loop and final end up into the centre of the seizing, can be allowed to protrude from the middle of the turns as otherwise it may be difficult to tug through (Fig 28).

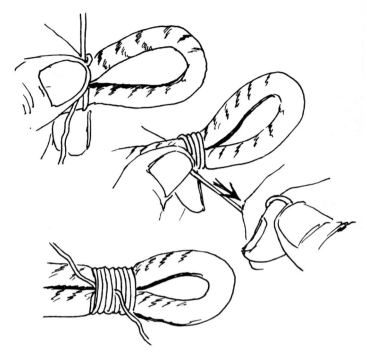

Fig 28. Seizing an eye in rubber cord.

A proper serving is laid on with a serving board (Fig 29). An eye splice in wire is first 'marled down' and taped over to render it smoothly tapered and the work is then set up taut between two strong points. At 3, the action of the serving board is shown and it is rotated *against* the lay of the wire or rope being served so that it is tightening the lay. While passing the board a gentle finger pressure is maintained on the marline or twine to make the turns tight and smooth. Once the end of the serving is reached, in this case the eye of the wire, the twine is led over the corner of the board (4) so that the turns can be laid on tightly right up to the limit. Having stretched the twine in this way, half a dozen turns must be slackened back so that a spike can be inserted allowing the end of the twine to be pulled home; the turns are then hauled tight again with the spike and finally the end is tugged home and trimmed.

While a serving board is necessary for natural fibre marline twine which is subject to contraction and expansion in wet/dry conditions, when using low or non-stretch synthetics, or indeed any synthetics that are unaffected by conditions aboard, it is less important. The

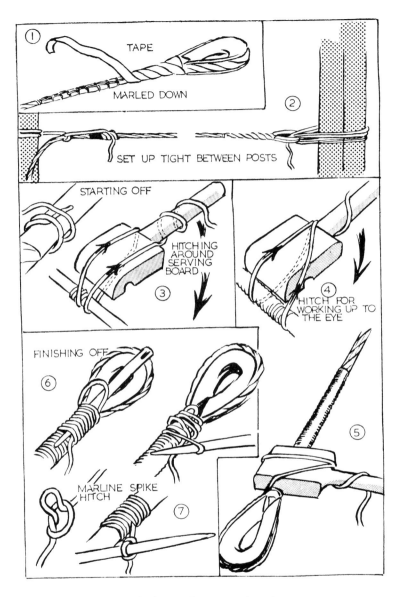

Fig 29. Proper serving laid on with a serving board.

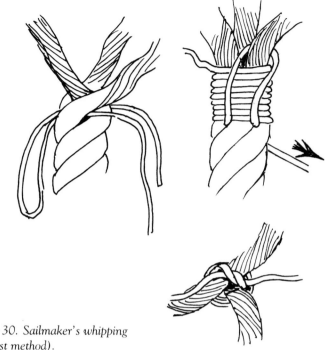

Fig 30. Sailmaker's whipping (first method).

aim then is to heave taut each turn as you make it. End off by overbinding the marline spike so that a tunnel is left through which to pass the end of the twine. Again, haul taut each of the slackened turns after removing the spike and snug the end with a sharp tug.

Whippings are largely superseded in small synthetic stuffs by fusing with a match or a lighter and it is both quicker and neater. In the case of synthetic rope, however, fusing calls for an electric cutter, which isn't part of the ship's bosun's bag. A whipping is then the answer. Don't buy those plastic rope terminals and don't back-splice a rope's end; both cause a knob on the end which can snag up and result in an emergency.

A simple whipping is no different to a seizing and, provided it is laid on tightly and against the lay of the rope in the case of three-strand rope, it is quite serviceable. A more secure method is the sailmaker's whipping.

There are two ways of making it. In the first method the rope is unlaid for a very few turns, the twine is doubled and laid in as shown, the rope laid up again, and the turns are laid on against the lay.

Finally, the doubled loop is carried up over its own strand end and tugged taut and then reef knotted to the other end of the twine right inside the heart of the three ends (Fig 30).

The second method is more secure than ever. The rope is not unlaid but the twine is laid on for the required number of turns, overbinding its end and against the lay (Fig 31). Using a sail needle, the other end is then sewn through the rope following the grooves between the strands, finally being forced right through a strand and trimmed off.

Fig 31. Sailmaker's whipping (second method).

The braidline eye splice

This double braided line is particularly suitable for small yachts as it can be used for either sheets or halyards, being very easy to handle and of low stretch. In passing it should be mentioned that this double braid rope, which has a braided core and braided outer cover, should be coiled by picking up a bight and placing the back of the right hand towards the palm of the left. In other words, no attempt should be made to form a normal coil. The bights thus made will form a series of figure-of-eights. If it is coiled clockwise in the normal manner, a series of turns will be introduced causing the rope to tangle later.

FITTING OUT

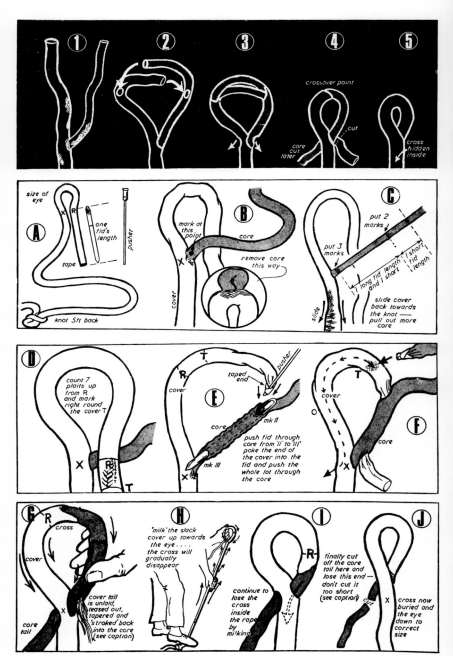

Fig 32. Braidline eye splice.

The double braid eye splice shown here retains about 85 per cent of the strength of new rope and the full strength of worn rope (on average). With enough practice it can be made complete in about five minutes. In principle, the core is removed from the cover and by clever tucking, each 'swallows' the other (see the black panel in Fig 32). The cross-over points are taken back into the standing part of the rope by holding in reserve a sufficient amount of slack outer cover. If a thimble is to be included in the eye it is very important to measure the size of the eye carefully; the thimble is then inserted before all the slack cover has been smoothed back. Thimbles are not strictly necessary. Old rope should be soaked in fresh water (and dried) to soften it. Put a knot in the rope 5 feet back from the end. Use a plastic fid and pusher (according to size of rope to be spliced), measure as shown, and mark X-R to give the size of eye required. Bend the rope (B) and part the threads of the cover at X to allow removal of the core. Milk the cover back towards the knot and then smooth it back into place again, just to ensure that the cover is lying at its correct tension; mark the spot where the core emerges. The cover can now be smoothed right back to the knot, allowing enough core to emerge for two new marks to be made, measured by the fid (see sketch C).

Now mark your cover (D). Count seven plaits up from point R towards the end of the rope and make mark T. Go back to your core (E) and push the fid in at mark II emerging at mark III. Tape the end of the cover to make a point and put it in the hollow end of the fid so that it can be passed through the core after the fid until mark T comes up against mark II. This is the cover tucked. Next (F) the core has to be tucked through the cover starting at T and re-emerging at X.

You now have something like sketch G. Pull out more of the cover until the tail that emerges can be tapered and then 'lost' inside again by stroking it back. Hold the cross-over point tightly and smooth away wrinkles to left and right. You are now ready to bury the cross-over (H). Place one foot on the knot made 5 feet back up the rope (first job remember) and begin 'milking' the rope gently at first, towards the eye which will gradually diminish in size. Tighten your grip by degrees. The cross-over *should* disappear inside the rope (1). If it bunches, give the remaining core tail a sharp tug to ease it. Finally, cut the remaining tail (J). If this is too short a flat part will result, but since the strain is divided equally between the two legs of the eye this will not weaken the rope unduly. Better to check stages A–H to see where, if at all, you have erred.

FITTING OUT

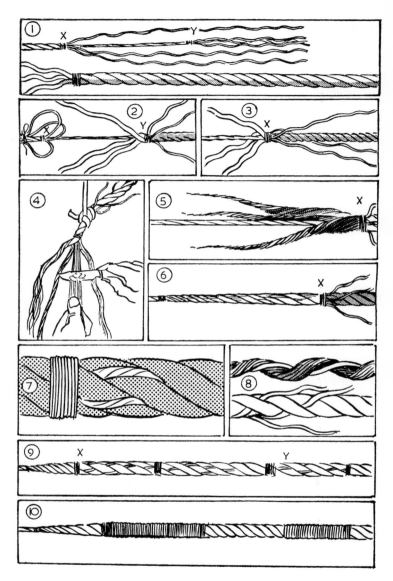

Fig 33. Tail splice.

Tail splice

There is one other splice sometimes needed in modern craft – this is the tail splice used for joining a rope tail to a wire halyard. Although the tail is intended only as a hauling part to bring the wire part down to the winch and subsequently to take the strain of a sail aloft, the splice must be very strong because there are times (the case of a very deep reef in the mainsail comes to mind) when the splice must take the whole load. Furthermore, unless well made, passage over a masthead sheave will soon cause it to fall apart. Once more there are two ways of going about it.

This is a difficult splice, although simple to understand. It is made difficult by the synthetic materials and by the stainless wire which is the normal material used nowadays; the characteristics of synthetic rope have been outlined, but stainless wire is very tough and springy in addition. The crux of the matter lies in preserving the natural shapes of the strands and wedding rope to wire in such a way that the sinuous strands of wire lie naturally along the lay of the rope.

Study the drawing (Fig 33) and unlay three of the six strands back to a point 'X' which is a stopping applied about 130 diameters back. Put another stopping about halfway back at 'Y' and unlay the remaining three strands. The wire will look like stage 1 in the sketch. Unlay the rope for the same distance as the wire back to a stopping at that point and 'marry' the three wire strands to the three rope strands at 'Y', Stage 2. (Marrying simply means putting them together so that the three rope ends each have a wire end between them.)

From this point the rope strands are laid up around the wire. To do this, each strand in turn is held by thumb and forefinger, given a twist clockwise to tighten the threads and laid, following its natural twist, around the wire so that it assumes its original shape but with the wire in the middle. Continue up to point 'X' and apply a stopping, Stage 3.

Now take a sharp knife and taper the rope's ends (Stage 4); do this exactly and evenly working down to a fine rat's tail, then wax each strand and proceed to lay them up, again with that back-twist, around the main body of the wire rope. The sketch shows how the work is suspended so that there is something to pull against while shaving down; Stages 5–6 show the completed taper.

The three wires at X must now be tucked, over and under, along the rope which you laid up around a wire middle, Stage 7. This is tricky; you must *not* kink the wire strands or bully them in any way. They must lie smoothly, conforming to their own wavy shape and the lay of the rope. There is a knack to this but it is impossible to describe adequately without an actual demonstration. Stage 8 shows how they must lie. Finally, tuck the wire strands from position Y in the same way, losing the ends in the heart of the rope and putting a serving over their graves; the job should now look like Stage 9.

The servings are applied after the splice has been stretched to make the parts settle. Rig it to a sheet winch or tackle. If you don't do this it will stretch in use and the servings will then become slack.

Tail splice in double plait

The tail splice in double plait consists of burying the wire into the middle of the rope and then splicing core and sheath into the wire. It is more complicated than this but not difficult to do. Needed is a special plastic fid supplied by the makers, British Ropes Limited, and suitable for the size of rope. I believe Samson ropes in the USA also supply this tool. It has a smooth point at one end and a hollow at the other; a 'pusher' (again supplied) or a long, thin electrician's screwdriver is also needed.

Tape the end of the rope (Fig 34) and make a mark one fid's length back (mark R), then measure a short fid's length – this is shown on the fid – and make mark X. Measure 5 fid lengths along the rope and tie a slip knot. Bend the rope at X, part the strands of the outer cover, and ease out the core; mark it at this point and call it '1'. Slide the cover back towards the slip knot to expose more core and, working along the newly exposed core, mark a short fid's length '2' and a whole length plus another shorty and make mark '3' (use three strokes of the marker).

Put a thin whipping on the end of the wire and mark back along it one fid length and apply some adhesive tape (dye-mark will rub off); call this W. This length is to be buried in the heart of the core. Now slide the point of the fid into the core at mark 2 and out again at mark 3. Insert the end of the wire in the hollow end of the fid and push the whole lot through until the 'W' on the wire is level with mark 2 on the core. Hold it firmly at this point and 'milk' the core towards mark 3 until the wire disappears.

The next job is to get the cover to swallow the core again. Remembering that you have slid the cover back towards the slip knot, you now have that amount of slack cover to 'milk' back again until mark X (where the core originally came out) and mark 1 have met. You should now be ready to tuck the two parts of the rope into the wire.

Using a hollow splicing fid for preference, lift three strands of wire, tuck the core through, then lift the remaining three strands and tuck the cover. You must then take each strand over three and under three. Give each part two full tucks, then make two half-tucks

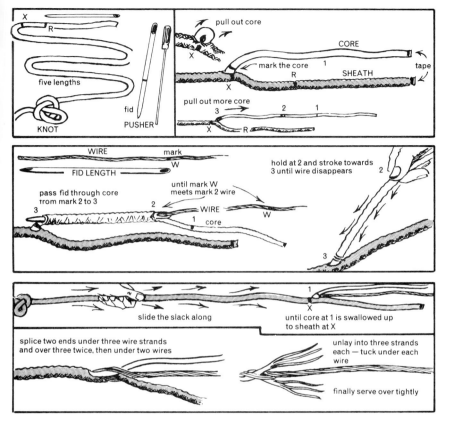

Fig 34. *Tail splice in double-plait.*

by tucking under two strands and finally two quarter-tucks. This is done by splitting the rope ends into three tails each and, lifting one wire strand at a time, tuck a tail under each. The ends are then shaved to a taper and the whole splice is stretched, on a sheet winch perhaps, and served over with adhesive tape. Serve tightly.

Racking seizing

Used to seize two parts of the line in parallel, usually at the eye of a pair of jib sheets, this seizing is designed to hold firm even though the strain is applied to one leg of the eye only (Fig 35). Make a slip knot around one leg of the eye and proceed to pass the strong twine in a figure-of-eight around both legs working towards the eye; heave each

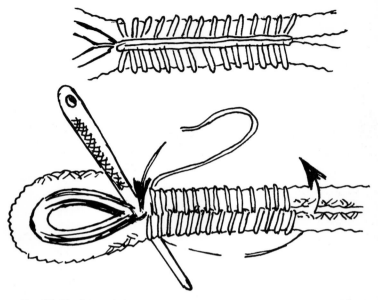

Fig 35. Racking seizing.

turn really tight. When the work has reached close to the toe of the eye take a fid and force a passage for the end of the twine (a large sail needle is a big help). The twine is then passed round and round the middle of the two parts, heaving tight each time, until the whole seizing is really tight. Finish by making a clove hitch around the middle and losing the end either inside or by passing it through the lay with the needle. As an alternative in plaited rope, use the sew-and-serve method.

Lacing to a spar

The spiral round and round lacing so often seen is quite wrong since it allows the sail to wander up and down the spar and because, should it part, the whole lacing slackens at once; a correctly laced line will not. Note that the lacing passes under its own standing part with each hitch so that it is trapped or nipped at each crossing (Fig 36).

Lanyards

Rigging set up tight by means of a lanyard is usually found in open boats, but older craft may also have, say, a fore topmast stay set up in

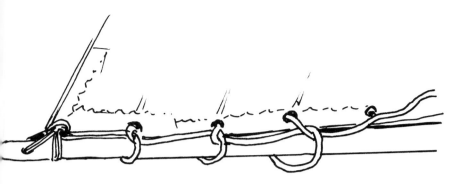

Fig 36. Lacing to a spar.

this way, and also bowsprit shrouds and guardrails. The head of a sail is hauled out along the gaff or the foot along its boom by means of a lanyard. In the latter cases the lanyard may be spliced through the hole in the end of the spar then doubled back through the sail cringle, through the spar again, and hauled taut to extend the sail. Further turns are taken around the spar and cringle to nip the sail close and the lanyard is finished off in a hitch around the hauling out turns.

Lanyards used to set up rigging either reeve through deadeyes or they require a chainplate with a flattened eye (Fig 37) and similarly a flattened or lanyard thimble in the end of the stay. If ordinary round thimbles and chainplate eyes are used, the turns of the lanyard will fall one upon the other. This will mean that the first two or three turns will carry the whole weight due to the fact that they are nipped. The remaining turns bear very little load until such a time as the whole system has had time to settle. A smear of tallow on rigging lanyards helps them to render easily. Finish off with two turns through the upper eye and several half hitches around the whole set. Lose the end by weaving it into the turns.

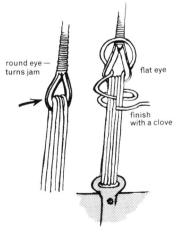

Fig 37. Lanyards.

A simple method of providing an emergency release for guardrail wires at the after pulpit rail. Stainless pins can be tugged out to release the hitch (see Fig 39).

The lower edge of a cockpit dodger can be held by means of rubber cord over hooks, thus enabling it to give underwater pressure in an emergency.

Dodgers

Most yachts are rigged with cockpit dodgers as a protection for those on watch and as a place to display the name; both are important functions despite the extra windage they cause. If they are badly rigged, though, these dodgers or name screens look quite terrible.

A dodger that is so securely lashed to the guardrails that it cannot be released without slashing it free with a knife is a potential menace to crew safety for two reasons. In the hopefully rare event of the yacht being at sea in extreme conditions of wind and wave, there is a real possibility of her being hurled flat on her side by a big wave. The weight of water that would be trapped by an immersed dodger could then prevent her from recovering before the next big sea struck her; the lower edge of a dodger that was either loose or releasable under pressure could then save the day.

The second threat lies in the risk of having a man-overboard emergency. The main problem in such an event is often less one of locating, returning and grabbing hold of the victim so much as one of getting him back on board. The cockpit being the best place for

FITTING OUT

rescuers to work from in terms of exerting their full strength, it is the guardrail and dodgers that then form the greatest obstacle; guardrails, or the wires at least, can be released provided the dodgers can be released with them.

The various release options were dealt with earlier, but the dodger needs separate consideration. A dodger must not be secured directly to the after pulpit eyes, but to the upper guardrail wire. Its lower edge should not be laced to eyelets along the toerail for the obvious reason, but to a series of open hooks or small thumb cleats; and rubber cord should be used instead of a lacing line. The forward end of the dodger, ideally, will be designed to come at a stanchion to which it can be seized top and bottom or laced but, failing this, an eye must be seized to the upper guardrail wire to which the upper forward corner of the dodger can be seized (Fig 38).

To make such an eye and secure it so that it will not slide along the shiny stainless wire under strain, proceed as follows: Tape a short

Roller headsail rotating line shown here secured by a simple hitch to a guardrail wire.

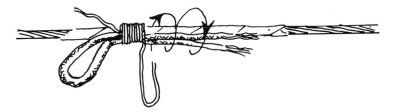

Fig 38. Seizing a dodger to the upper guardrail.

length of the wire to provide an anchorage. Take a length of signal halyard or small three-stranded stuff, double it to form an eye and seize it tightly, working forward from the eye. A clumsy but effective alternative is to use a small bulldog grip, lash to it, and tape over all.

Rubber cord

Bungee, shockcord, or whatever it may be called, becomes, like self-adhesive plastic tape, one of the primary standbys of the modern yacht. Rubber cord has a dozen uses both on deck and below and seizing an eye in it (mentioned earlier) is about the only technique involved. Never buy cheap rubber cord; not only will it be less powerful in terms of elasticity, but it will quickly become even weaker and it starts to perish from the moment it is exposed to sun and elements. It is possible to buy various end terminals that save having to seize eyes and the like, but beware of injuries. In particular, take care with the plastic balls used in making sail tiers or tyers – these can flip free during use and knock out a tooth or enter an eye. The best of all sail tiers is a 2 metre (6 feet) length of sailcloth run up cheaply by your sailmaker; a pair of these, or maybe three to an average mainsail, is all you need.

Taping split pins

That all open pins should be taped is common sense; not only can exposed pins tear sails but they can gash hands too. Try not to touch the sticky side of tape with the fingers; wind it on tightly and cut with a knife at an angle. Properly used, self-amalgamating tape is better since it welds itself into a solid mass. This tape must be stretched while it is being wound on – the effect being to bring about some chemical change which causes the layers to 'amalgamate'. Finish by overstretching as if intending to snap the tape, and cut while under tension.

Quick release dinghy lashing

The arrangement in Fig 39 is probably self-explanatory. The lashing consists of two lengths of rubber cord which are strained together with a temporary lashing pending the rigging of the release line. This

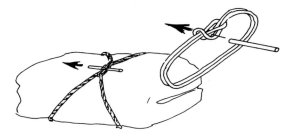

Fig 39. Quick release dinghy lashing.

is a grommet or endless circlet. It is passed through the two rubber lashings in the manner shown and the release peg is inserted. Finally, the temporary lashing is cut and the grommet and peg are left holding the strain. This arrangement is excellent if it is intended to carry a half-inflated dinghy on deck.

Ground tackle

Ordinarily one hardly even thinks about anchors and cables when fitting out; there isn't much to go wrong with them provided they are made by reputable people. In the case of the very light versions of welded anchors, Danforth and Meon, there is always a possibility that the shank may become bent out of line by being dragged out of a heavy clay seabed. Very rarely this may also happen to larger anchors which have been subjected to a massive lateral pull – perhaps following a wind shift during gale conditions but only if the anchor is lodged immovably on the seabed, wedged between rocks perhaps.

A chain anchor cable should be end-for-ended every year or so because the inboard end rarely does any work, remaining in the chain locker 90 per cent of the time. If the cable is made up from two or more separate lengths of chain with joining shackles between each, these should be checked over carefully. If they are to achieve their proper function, these joining shackles, whatever type they may happen to be, should be removable in an emergency. With an anchor fouled on some seabed obstruction the option of being able to slip and buoy an anchor simply by disengaging a shackle avoids

Bitter-ending the cable by means of a rope's end leading to a strong point below allows the whole cable to be hauled up on deck and absorbs accumulative twists in the chain.

having to slip the entire length. Shackles may be split link or pin-and-pellet, both needing to be parted with a hammer and punch. Carry an iron or lead block with a hole in it as an anvil, and also carry spare shackles.

Most anchor cables are either a single length of chain or a combination of chain and rope, but in either case the cable must be marked at intervals so that crew can see how much scope has been veered when anchoring. Five fathom or 5 metre lengths are usual and marks may be painted (eg one red length at 5 fathoms, two at 10 and so forth). Or short lengths of signal halyard can be used – one tail, two tails, etc.

Rope/chain splice

Although it is quite common to see a 5 fathom length of chain next to the anchor joined to the rope cable by means of an ordinary 'D' shackle and a spliced eye, it is bad practice. More often than not, in trying to find a shackle that will fit the chain link, a size that is far too

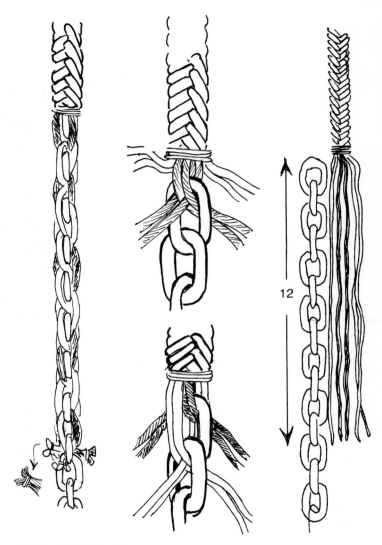

Fig 40. Multiplait splice.

small is used – thereby producing a weak spot in the cable as a whole. Or the spliced eye is so large that it cannot pass through the pipe that takes it below decks.

The correct method is the rope/chain splice. When the rope component of the cable is of Multiplait construction – and it is without doubt the best of all ropes for an anchor cable – the distinctive

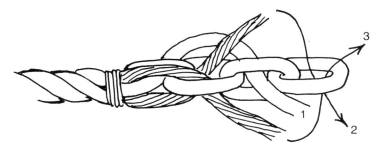

Fig 41. Three-strand splice.

Multiplait splice is used (Fig 40). In the case of ordinary three-strand rope, the matter is a little different. The method usually employed for attaching chain to rope uses only two of the three strands and therefore creates a strength loss; the third strand is cut off short and overhand knotted to one of the other returning strands as in a long splice. The splice shown (Fig 41), although slightly bulkier, uses all three strands.

Having unlaid a suitable length of rope (about 24 diameters) and whipped it to prevent further unravelling, pass two strands from opposite directions through the *second* link of the chain and pass the third strand through the end link; there are two ways of proceeding from here. Either the strands can continue to dodge in and out of the links from opposite directions with the third strand following back and forth, finally over-serving the three ends very tightly, or the strands can simply be spliced back up the rope in the ordinary way (Fig 42). The latter method may be a little bulkier, although tapering off the ends of the strands for the final tucks minimises this. There is a 'loose' area between rope end and chain which needs serving over. Remember to bind the ends of each strand with Scotch tape right to the tips.

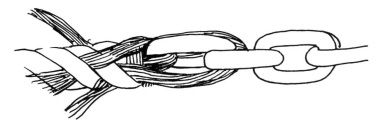

Fig 42. Splicing back up the rope.

12

Sail repairs

The extent of sail repairs attempted will depend upon the condition of the sails, the owner's skill and the urgency. A set of sails in first-class condition but damaged in some way can easily be ruined by an inexpert amateur. For instance, a patch put on at the wrong tension will throw a set of strains in several directions and ultimately the sail will tear again, perhaps more drastically. The shape of a sail can also be spoiled and a bodged-up seam can be mangled to the extent that it is hard even for a sailmaker to rectify it later.

The justification for tackling one's own repairs may be that the sails are old and greatly patched (safety dictates that they should be replaced anyway) or that the sailmaker is too busy to cope with the spring rush. Sail repairs should not be left until the spring. Get-you-home repairs may often be a matter of using adhesive tape (sold for this purpose), but in heavy weather stronger, albeit clumsy, work may be necessary.

Machine-sewn Terylene sails are difficult to repair neatly by hand because the machine will have produced a flat seam at even tension with the zig-zag stitches lying on the surface of the cloth, and hand sewing or seaming involves much tugging which produces a different tension and tends to bury the stitches. In fact, the main reason why the stitching of a machined Terylene sail is so vulnerable to chafe and wear is that it is not tugged home deeply. Hand sewing, which does just this, raises little mounds between each separate stitch which protects them.

With several feet of seam to re-sew, lay the sail flat and position the seam, then make pencil ticks or 'strike-up' marks across the seam as a guide to tension; these ticks must be in line when the sewing is completed (Fig 43). To aid this, a fine needle or safety pin should be put through the two thicknesses every 6 inches or so and removed in turn as they are reached. Choose a needle the greatest thickness of

SAIL REPAIRS

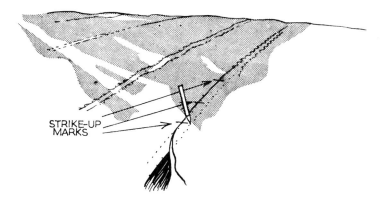

Fig 43. Make pencil ticks as a guide to tension.

which will allow the threaded eye to pass through the hole made by the needle's triangular section. Choose the smallest needle consistent with carrying a thread of suitable size.

Using waxed Terylene thread, over-sew the end and proceed as shown in flat stitch (Fig 44). There is a knack in using a sailmaker's

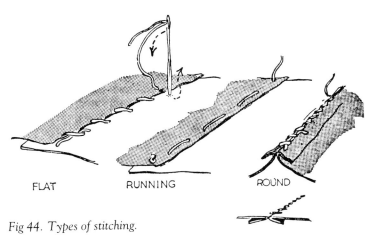

Fig 44. Types of stitching.

palm (Fig 45). Note that the finger and thumb 'learn' to return the eye of the needle to the iron of the palm after each stitch. The second sketch shows the hand ready to push the needle, and lastly how the second finger is steering the eye back to the iron. Lay a thick piece of canvas over the knees before starting work otherwise you will jab yourself. Ideally the fingers of the sewing hand go flat as the needle goes in but at first the wobble will be hard to control.

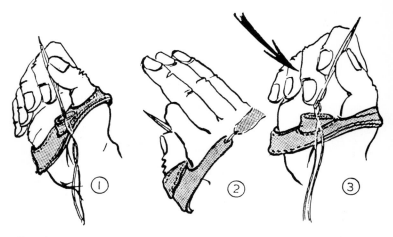

Fig 45. Using palm and needle.

Try to work at eight stitches to the inch, but if you can pick up existing holes in the seam so much the better. Needles range in size by numbers. A 15 is a good average size, but 16–17 for light sails and a 12 for odd heavy jobs is usually enough. They must be stored in oil-soaked cloth to prevent rust as a rusty needle is almost impossible to use. Spinnakers are extremely difficult to hand sew. Forget the sail needle and the palm and either let the professionals do the job or ask a dressmaker for help. The tension must be exactly right otherwise the sail is likely to tear again the next time it goes up.

Other stitches shown here (Fig 44) are the running and the round stitch. The former can be used either to tack a patch in place (as in dressmaking) or for sewing very thick or multiple layers of cloth – the running stitch can return along itself, using the same holes from the opposite direction and filling the gaps. The round stitch is suitable for sewing up covers, ditty bags and the like. It is easy to do and, when the work is finally flattened, makes a tight seam.

Patching

Lay the work flat and place a patch as shown (Fig 46) over it, then pin it in place; sometimes an office stapler can be used and the staples removed later. Use old sailcloth on an old sail which has done all its stretching, otherwise the patch will stretch and in time become slack. This is more apparent with natural fibre cloth than with synthetics however.

With the patch firmly pinned in place turn under its edges after

SAIL REPAIRS

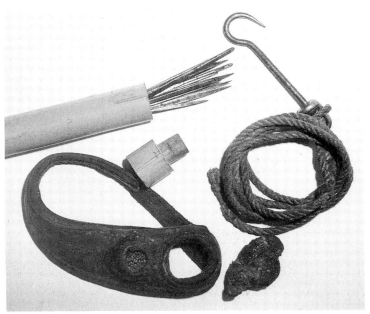

Sewing palm, needles, bee's wax and sail-hook. A good way to ensure a snug fit is to soak the leather palm in water for an hour and then wear it until it is almost dry.

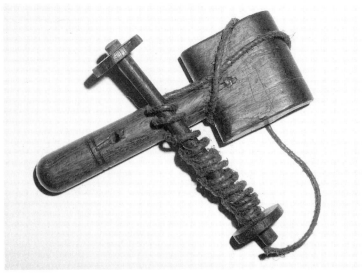

Serving board and reel.

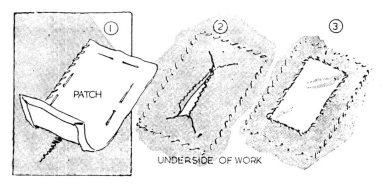

Fig 46. Patching.

mitring the corners and stitch it, using a flat stitch. Turn the work over and treat the tear as shown (Fig 46[3]), turning the edges under and again flat seaming all round. Big L-shaped tears are more difficult to deal with and it may be necessary to lay a patch over the whole width of a panel or cloth.

Darning

It is difficult to make a neat darn. Heavy cotton sails can be darned neatly enough but the light Terylene sails of today are a problem. Use the lightest yarn practicable with a needle to match so that the line of holes is less damaging. Use of a big needle loaded with heavy yarn merely results in a clumsy row of punctures. The sketch in Fig 47 shows the method used, but in the case of old sailcloth, especially cotton cloth, the stitch holes should be staggered and set well back away from the tear.

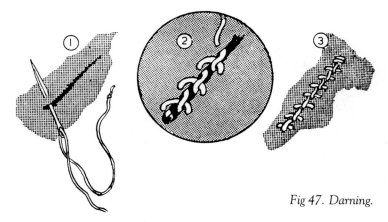

Fig 47. Darning.

SAIL REPAIRS

Batten pockets

One of the more usual repairs is to the end of a batten pocket and occasionally to its open mouth. Stitch wear is commonly the trouble and this is a case for following existing holes by hand. Batten pockets, which form a hard spot in a sail, take more punishment against rigging than other parts and need to be particularly strong, but the leech of the sail is the critical thing. If the leech should part, the sail is liable to tear straight across. When repairing a pocket pay close attention to the state of the leech and its stitching.

Roping

Sails are roped on their port hand side and, as the stitching is well buried, it is unlikely to fail on Terylene sails (Fig 48). The tension of the sailcloth along the rope is critical to the shape of the sail when set and no repairs other than renewal of stitching should be attempted unless the risk of a change of shape is acceptable.

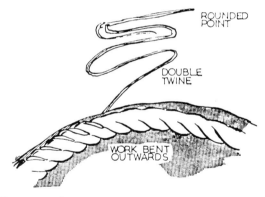

Fig 48. Roping a sail.

Eyelets and cringles

The brass ring and grommet eyelet are completely simple to fit as long as you have the punch and dolly to go with them. The hole can be cut with a sharpened tube and the eyelet assembled through it. Lacking the proper tools, a fair job can be made of it with a ball-bearing which is tapped down to open the brass neck of the grommet.

Sewn eyelets are the alternative (Fig 49). Make a hole rather smaller than you'll need and widen it with a fid (1). A galvanised

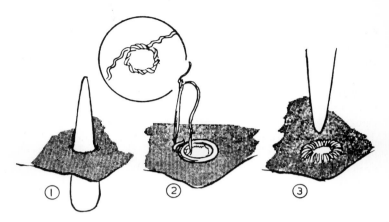

Fig 49. Making a sewn eyelet.

steel ring is then stitched in neatly and the hole finally trued up with the fid again. Lacking the steel ring, a rope grommet can be made from a single strand laid up on itself as seen in the inset.

Cringles for tack, clew or reef earrings are made in a similar way (Fig 50). The strand, which will be just over three times the length of the finished cringle, is passed through one eyelet, laid up around itself, first with one strand until you reach the opposite eyelet, and

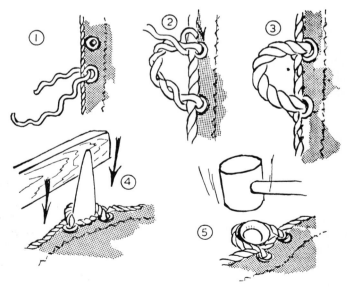

Fig 50. Making cringles.

then back with the other end of the strand forming the three parts of the 'rope'. A fid is then forced through the cringle to stretch it wide enough for a brass thimble to be dropped in. If the size of the rope cringle has been gauged right it will be just possible to stretch it far enough for the thimble to be knocked in – after that it should contract at once and hold the thimble firmly for ever. This use of a brass thimble also applies to those big eyes which are worked in the sail as described when dealing with smaller eyelets earlier. The method is just the same including the steel ring (galvanised), but in order to protect the stitching a brass thimble is slipped through and then belled out all round with a hammer.

Reef points

These are simple to replace; each nettle or reef point passes through a small eyelet in the sail and is middled there and held by stitching each side through the sail to the other side of the point.

Modern synthetics

The preceding notes are mainly concerned with traditionally built sails, and in fact many of the sails to be found in today's yachts are definitely best left alone by the amateur. This applies particularly to lightweight sails such as spinnakers and cruiser chutes which would be ruined by palm-and-needle repairs. The adhesive tape repair kits available should be an essential part of the seagoing inventory, as should small pieces of sailcloth of various weights and the fabric adhesive for glueing on temporary patches. The aim of repairs made while cruising is to prevent damage from worsening and allowing the sail to be used pending a visit to a sailmaker.

Cleaning sails

The cost of having sails washed annually by a sailmaker is not excessive, particularly as he can be asked to check over the sail at the same time. Don't send sails to an ordinary laundry; they may come back looking clean but they may also come back badly damaged in the process.

If you decide to clean your own sails, don't try putting large sails into the family bath; not only will it be marked by sail hanks but the whole job is confined and messy – especially when it is time to carry the whole dripping heap outside into the garden. I prefer to fill my inflatable dinghy with water (having rinsed it out first with soapy water) and use it as a laundry. The hosepipe can then be used for final rinsing. It is often possible to drip-dry a sail by hoisting one corner to a bedroom window and spreading it out by means of a rope attached to a garden fence or fruit tree.

Stains such as rust marks and mildew often yield to lemon juice and neat detergent. In severe cases, a mild solution of oxalic acid can be applied and then rinsed off immediately; rinse several times. White spirit shifts most oil stains but avoid household bleaches and strong chemical cleaners – better a dirty patch than a weak patch.

Ropes

An annual twenty-four hour fresh-water soak followed by drip-drying will remove all salt, for it is salt that stiffens ropes and increases internal friction. If a rope is oily or dirty, wash it in warm soap suds followed by a fabric softener, then rinse well and drip-dry.

13

Safety

Certain aspects of safety have been covered in earlier chapters (eg the bilge pump, non-slip treatment of decks, and guardrails and stanchions). However, specific items of safety equipment merit separate coverage.

Jackstay and safety harness

The provision of a permanent jackstay along both sides of the deck running from cockpit to bows is an essential part of harness use, which becomes quite pointless unless the wearer has adequate hook-on provision. The usual jackstay arrangement consists of a pair of purpose-made flexible stainless steel wires shackled to eyes in the outside cockpit coamings and running to other strong-points on the foredeck. If a yacht is rigged with a babystay the jackstay can be an unbroken single wire leading close around the babystay. This means that a harness wearer can hook on to the jackstay on, say, the port sidedeck, go right forward, around the babystay, and back to the cockpit via the starboard sidedeck without having to unhook. In some cases the hook rope scope even allows access to the bows of the yacht (Fig 51).

Since a jackstay lies on the deck it can form a hazard in its own right because it can roll under the foot of an unwary crew; for this reason, it is wise to run the wires close to the sides of the coachroof rather than on the deck. Jackstays do not need to be wires and any suitable size of low or prestretched rope can be used. It is equally efficient and probably less of a hazard underfoot; however, if rope is used it should be disconnected and stored during the winter.

Safety harnesses should be examined carefully for signs of defective stitching and washed thoroughly to remove salt – a stiff harness is hard to adjust to fit people of different girths or to wear over extra clothes and oilskins. It is because they are so hard to adjust that harnesses are so often worn too slack for real safety.

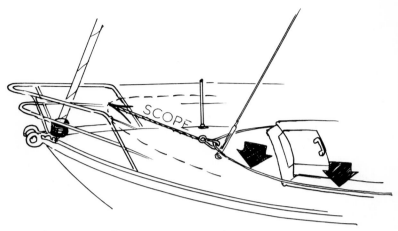

Fig 51. *Harness allows wearer access to bows.*

Lifejackets and buoyancy aids

All should be washed and thoroughly dried; in the case of inflatable jackets, these can be checked for air leaks while washing and rinsing – if none are found they should be fully inflated and left for at least twenty-four hours. The inflation valve is quite commonly sticky from lack of use but it is easy to free or to replace. Any patching is best left to a manufacturer's overhaul. Buoyancy aids are usually built with pads of closed cell plastic foam which needs no attention.

Zip fasteners and poppers are a notorious weakness of all such safety garments, and if a lifejacket or aid cannot be zipped up tight it will become dangerously inefficient. Fresh-water washing may free a salt-choked zip, but it is more likely to seize up completely. Apply penetrating fluid and leave overnight. If the zip still won't shift, scrape away as much of the corrosion as possible with the point of a fine knife and apply more fluid, working the zip with the fingers. Several applications may be needed but never force the zip. Once freed, brush the teeth with a fine wire brush and apply zip-free (a solid-stick lubricant usually available at haberdashers). Poppers can be similarly treated and lightly oiled. Check safety lights on life-jackets, whistles, reflective patches, etc.

Distress flares and rockets

Although the widespread use of VHF radio in an assistance calling role has done a lot to distract attention from the importance of

The safety harness jackstay which runs the length of the deck both port and starboard is a basic safety essential. The one shown here is slack to permit the anchor to be lifted from beneath it.

distress pyrotechnics, they can still be vital in particular circumstances – such as the failure of yacht electrics, the inaccessibility of the VHF set in a sinking boat, the loss of the aerial by dismasting, and of course their use in guiding rescuers to the scene. At fitting-out time these pyrotechnics must be examined for condition and dryness and checked for date. Out-of-date rockets and flares should be taken to the nearest Coastguard station and handed in for controlled destruction. *Do not* save them to use as Bonfire Night fireworks – old rockets in particular can become erratic and there is one case on record of a careless handler being chased round his own kitchen by a runaway rocket.

The magazine *Yachting Monthly* once invited readers to hand in their old pyrotechnics so that they could be tested on the manufacturer's proving range. The majority worked, but performance varied from erratic and dangerous to low altitude, loss of colour, and falling to earth while still burning fiercely. Hand-held flares discharged in clamps threw out quantities of molten red-hot dross (a disaster in a liferaft!). Pyrotechnics that have only just passed their expiry date can, of course, be kept on board as a back-up for newly purchased ones, but they should not be kept longer than a couple of years.

Fire extinguishers

As in the case of distress signals, fire extinguishers have an expiry date after which they must be given a maker's check and overhaul. *Yachting Monthly* carried out a similar operation with reader-owned fire extinguishers of many different types and makes, sixty-five in all. Over 30 per cent of these either failed to work at all or proved to be faulty in some way, particularly those that were well overdue for a maker's check.

A common fault, irrespective of performance, was that extinguishers were far too small for the role they were expected to play in the size of boat they came from. In general terms, a 30-feet (9-metre) yacht having an engine and galley should have not less than two 3-pound (1.4-kilo) dry powder extinguishers or their equivalent in capacity and effective coverage. A larger vessel calls for a minimum of two 5-pound (2.3-kilo) extinguishers or/and an automatic engine-room system.

Suspect very old extinguishers, which may have lost weight or propellent gas. In the case of powder types, give them a vigorous shake-up to loosen any caking of the contents.

Liferaft and dinghy tender

A liferaft annual examination certificate may be costly to maintain, but it is important both for peace of mind and to the resale value of an expensive piece of equipment. Consider your liferaft in terms of suitability for you and your crew. As owners change boats and as crews either become more numerous or decline in numbers, the liferaft must reflect the current situation. The raft that was adequate for two adults and two small children can become inadequate when children metamorphose into hefty teenagers.

While a dinghy cannot be considered a viable substitute for a raft (with the exception of specialised dinghy/raft designs), it should be remembered that not all emergency abandon-ship situations take place at the height of a storm. A yacht may be sinking rapidly after hitting a rock or floating debris, or she may be on fire. It is then a question of how many people the dinghy can carry safely. Clearly the massive buoyancy of the inflatable dinghy scores highly. However, it is not merely a matter of how many people can be kept afloat by it, but for how long and with what mobility for reaching safety. Modern yachts accommodate a much greater number of crew than did earlier yachts of comparable size and the inflatable should take account of this.

Where a hard dinghy is carried it is seldom that one which offers total crew capacity can be carried on deck. In fitting-out terms this problem comes up for review – a bigger dinghy or a second back-up inflatable, which is the better compromise? In the case of hard dinghies, wooden or GRP, the matter of fixed buoyancy is important.

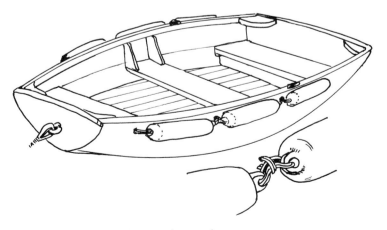

Fig 52. Small fenders also provide extra buoyancy.

Here we have a problem of fitting extra buoyancy without reducing internal passenger room. It is now possible to buy inflatable fender-collars which surround the boat at gunwale height. Alternatively, the permanent lashing of small inflatable fenders necklace fashion along the outside of the gunwale provides both extra buoyancy and protection (Fig 52).

Lifebuoys and danbuoys

Although there is little that can go wrong with these items, they can deteriorate from long exposure. Examine closely for chafe and UV damage, and check the mode of stowage and instant release capability. Orange Day-Glo paint fades rapidly without a crew being aware of the fact; touch up such paint annually. Check auto water-operated lifelights which are notorious for failures, and check battery-operated equipment.

Retro-reflective adhesive tape can be bought at most chandlers. Put it on lifebuoys, lifejackets, danbuoy staffs and the crew's oilskins. Having done so, provide a cockpit hand torch of high power and devise a permanent night-stowage position for it.

Retro-reflective tape applied to the staff of a danbuoy. Manufactured by 3-M, it has many safety functions – including use on lifebuoys and lifejackets.

Index

alloy mast, 54
amalgamating tape, 107
anchors, 108
anodised spars, 54
antifouling, propeller, 46
antifouling paints, 43
anti-freeze, 81

baby-stay, 65, 121
bacteria, 71
battens, 117
batteries, 75
beadings, 20
bee's wax stopping, 55
bilge pumps, 73
blisters (GRP), 30
boot-topping, 44
bosun's chair, 56
box mast, 56
braidline eye splice, 95
buffing, 32
burning off, 38

canvasing decks, 16
cap shrouds, 63, 65
caulking, 16
chain cable, 108
cleaning brushes, 42
cockpit lockers, 77
colour (gel coat), 24
condensation (tanks), 81
control cables, 81
corrosion, 70
cove lines, 48
crazing, 30
cringles, 117, 118
crosstrees, 63
curing times, 28
cutting in, 50

darning sails, 116
deadeyes, 103
deck, 18
  cavasing, 16, 17
  leaks, 9, 76
  mastic, 17
  plastic covered, 18
  seams, 9, 16
delamination, 32
dezincifying, 83
diesel engine, 80
dinghy, 125
distress flares, 123
dodgers, 104

electrics, 75
electro chemical action, 54
electrolysis, 10
engines, 80
eyelets, 117
eyesplice, 88

fastenings, 14
fid, 67
fillers, 40
fillets, 12
fire extinguisher, 124
five-part tackle, 58
flemish eye, 88
freshwater tanks, 71
  filters, 71
fuel, 81
  tanks, 8
fuses, 75

gaff rig, 67
galley pump, 74
gantline hitch, 60
gel coat, 23

INDEX

glue, 15
going aloft, 56
gouges (in GRP), 31
grain filler, 40
graving pieces, 12
gribble worm, 44
grommet, 108, 117
ground tackle, 108
guardrails, 77, 104

halyard
  tailing new, 56
  winch, 58
hand varnishing, 55
harness, safety, 121
hatches, 76
holding tanks, 73

impact screw remover, 84
internal halyards, 61

jackstay, 121
joining shackles, 108
jubilee clips, 69

keel
  bolts, 10
  moulded, 23
keying surface, 34
knock-down, 79

lacing, 103
lanyards, 103
lifebuoys, 126
lifejackets, 123
liferaft, 125
lights (electric) 75

man overboard, 105
marling down, 93
marrying, 99
masking tape, 50
mast
  rake, 65
  sheaves, 61
matt (GRP), 33

messenger lines, 52
metallic primer, 39
monel wire, 67
mouse, 61
multiplait, 110

name dodger, 78, 104
needles (sail), 113

one-pot system, 35
osmosis, 26
oxalic acid, 25

paint
  brushes, 44
  stripping, 38
painting technique (GRP), 36
palette knife, 41
palm (serving), 113
patching sails, 116
pawl springs, 74
petrol engines, 81
plywood-built, 11
polyurethane paint, 41
pole spar, 55
pricker testing, 10
primer (GRP), 34
pump impeller, 81
pumps, 72

quick release, 78

racking seizing, 101
ratstail, 99
reflective tape, 126
ricker spar, 55
rigging, setting up, 63
robins, 52
rope-chain splice, 99
roping sails, 117
rubber cord, 107

sacrifical anode, 35
sail
  cleaning, 119
  repairs, 112

sailmaker's whipping, 94
scraping, 45
scratches (GRP), 29
seacocks, 69; plugging, 8
seizing, 87
self-amalgamating tape, 48
serving, 87
set flying, 67
setting up (rigging), 63
sew and serve, 88, 90
shackles, 87
shakes, 55
shores, 44
shuffling time, 15
siphoning, 70
snake, 61
splicing, 87
spontaneous combustion, 26
sprung spar, 56
stainless steel wire, 52
stains (hull), 25
stanchions, 77
standing rigging, 52
step, 62
stenhouse slips, 78
stoppers, 40
stress crazing, 22
strum box, 75

tack rag, 43
tail splice, 98, 100
taper plug valves, 70
teak oil varnish, 47
teakwork, 26
teredo worm, 44
toilets, 72
tools (bosun's), 86
trowel cement, 37
two-stroke engines, 82

varnishing, 45

waterlines, 48
wet edge, 36
wheel valves, 70
whipping, 87
winches, 74
windows, 76
wood
  rot, 10
  screws, 83
wooden spars, 55

X-ray (keel-bolts), 11

zip fasteners, 123